GCSE Chemistry

Making a success of your coursework, revision and exams

Studymates

25 Key Topics in Business Studies
25 Key Topics in Human Resources
25 Key Topics in Marketing
Accident & Emergency Nursing
Business Organisation
Cultural Studies
English Legal System
European Reformation
GCSE Chemistry
GCSE English
GCSE History: Schools History Project
GCSE Sciences
Genetics
Land Law
Macroeconomics
Memory
Organic Chemistry
Practical Drama & Theatre Arts
Revolutionary Conflicts
Social Anthropology
Social Statistics
Speaking Better French
Speaking English
Studying Chaucer
Studying History
Studying Literature
Studying Poetry
Understanding Maths
Using Information Technology

Many other titles in preparation

Studymates

GCSE Chemistry

Making a success of your coursework, revision and exams

David Howard

BSc(London) GradIPM DipCG

www.studymates.co.uk

ISBN: 1-84285-008-3

First published in 2001 by Studymates Limited, PO Box 2, Taunton, Somerset, TA3 3YE, United Kingdom.

Telephone: (01823) 432002
Fax: (01823) 430097
Website: http://www.studymates.co.uk

Important note: Chemical experiments may be dangerous and should only be carried out under the supervision of a qualified teacher.

Typeset by PDQ Typesetting, Newcastle-under-Lyme, Staffordshire
Printed and bound in Great Britain by the Baskerville Press Ltd, Salisbury.

Contents

List of illustrations

Figures

Tables

Preface

This book has been written to help students following most of the GCSE Chemistry syllabuses. There is now a fair degree of uniformity in both content and approach between the various syllabuses. Care has been taken to cover all the major GCSE requirements. The book should also help you if you are following other courses such as Nuffield Chemistry, with its experimental approach. It is also intended to stimulate interest in this most absorbing and important branch of science.

The main idea of the book is to stress the *principles* of the subject. By improving your understanding of these principles, you will reduce the amount of material you need to memorise, and gain a better understanding of the subject as a whole.

There is quite a strong emphasis on an experimental approach – chemistry is after all a practical subject, with many practical applications in modern life.

The text is presented in a logical sequence, the chapters forming the building blocks of the whole. The chapters follow a regular format. You will find a handy one-minute summary at the start of each chapter, and helpful learning material at the end. This is made up of progress questions, practical assignments, discussion points, and study and examination tips, all carefully designed to consolidate your learning and understanding.

This new Studymate should really help you succeed on your all-important GCSE Chemistry course.

Good luck with your studies.

David Howard

davidhoward@studymates.co.uk

1

What is Chemistry?

One-minute summary – Chemistry is the branch of science that studies the chemical elements, the compounds they form, and the chemical properties of both. In this chapter we also discuss the importance of chemistry, and its part in our lives. A key principle is that, in any chemical change, **new substances** are formed. Generally the changes are difficult to reverse. They are often accompanied by energy changes, for example heat and light are given out. For example, fireworks contain combustible chemicals that react together when ignited, giving out heat and light. A simple example of a physical change is a piece of ice melting to form water; chemically speaking a new substance has not been formed. Central to the whole subject is the Periodic Table of the elements and their properties. In this chapter we will explore:

- the role of chemistry in our lives
- the basic principles of chemistry
- mixtures and compounds separation and purification.

The role of chemistry in our lives

Chemistry is enormously important in our everyday lives. Think of some of the common chemicals we use every day – sugar, water, salt and household bleach. These substances, and many others too, are often manufactured on a large scale. The starting point is often a mining operation.

The chemical industry

The chemical industry is an important part of our society. Examples of substances produced on a large scale are:

sulphuric acid
bleach
paints
plastics
fertilisers.

Some substances, such as salt, can be obtained from the ground in an impure state, but have to be purified for household use.

- The term substance in chemistry can mean a single species, an element, or a pure chemical compound.

The basic principles of chemistry

The basic principles of chemistry cover how and why chemical substances react together, both quantitatively and qualitatively.

Chemical reactions

The type and range of chemical reactions is very large, but many of them follow a pattern. Once you understand the principles, there is less you will need to remember. Here are some simple, but important, examples:

1. Metals and non-metals often react together in the process of compound formation.

2. Acids and alkalis always react together. Water is always formed in the reaction.

3. When elements enter into chemical combination they lose, gain, or share electrons with whatever other substance they are reacting. The other substance can be another element, or a compound.

All these chemical reactions will be covered in more detail later.

Other principles of chemistry

Other important key principles of the subject are:

(a) the Periodic Table of the elements

(b) chemical formulae and equations

(c) bonding in elements and compounds, and the transfer of electrons during chemical reactions

(d) oxidation and reduction.

Understanding the principles

Once you have understood these principles, most of the rest of the subject will 'fall into place'. The amount of factual material that you will need to learn by heart will then be much less.

Our starting point is to look at the differences in chemistry between mixtures and compounds. We will then describe how to separate the components of a mixture, and also ways of purifying solids and liquids. Then in chapter 2 we discuss the various particles which make up all matter and are the constituents of all the elements.

► Key point – Chemistry is a practical subject. We can explore a good deal of it experimentally.

Mixtures and compounds – separation and purification

What is a compound?

A compound in chemistry is a single substance made up of two or more elements. It is not usually very easy to split up compounds into their component elements. Water is a simple example. It is a compound, but not easily split up.

What is a mixture?

A mixture on the other hand consists of two or more substances, not combined with each other. Mixtures are normally fairly easy to separate. Air and sea water are both examples of mixtures. Let's look at how to separate mixtures.

Separation methods

Choosing a method for separating the components of a mixture depends on the properties of the substances involved. However, it is not always easy to remove small amounts of impurities from a substance. Therefore, in the synthesis of many industrial chemicals, the best way is to start with pure materials. The manufacture of concentrated sulphuric acid is an example.

- *Separating a solid from a liquid* – The process of separating a solid from a liquid is called **filtration**. The mixture is steadily poured through a filter paper in a funnel with a beaker underneath. The solid is retained by the filter paper, while the liquid passes through.

- *Obtaining a solid from a solution* – The solution is heated in a suitable container (an evaporating dish is often used). The heat drives off the solvent. We can evaporate the solution down to dryness by driving off all the solvent, or concentrate it by the same process and then leave it to crystallise out.

Crystallisation

Crystallisation is particularly suitable for purifying solid substances. The substance is dissolved in hot solvent, usually water for salts and ethanol for organic substances. It is then allowed to cool quickly to give small crystals. The reason for this is that small crystals are more pure than big ones. There is always some loss of the substance, called the **solute**, in the process.

Paper chromatography

Paper chromatography is a way of separating two or more liquids mixed together. It is based on the fact that the components of the mixture are absorbed at different rates by soft absorbent paper, such as filter paper or blotting paper.

- *Method* – Small quantities of the mixture are dropped onto the centre of a large piece of filter paper. The process is repeated several times, and eventually a series of different coloured circles are obtained. The most soluble substance is the furthest from the centre. The end result is referred to as a **chromatogram.**

Separating liquids

1. If the two liquids do not mix, then separation is carried out using a tap funnel. This is a fairly long, straight-sided funnel with a tap at the bottom (figure 11, chapter 8). The mixture to be separated is poured into the funnel. When it has settled into two definite layers, the tap is opened and the lower liquid passes out of the funnel into a suitable container. The lighter liquid, which was on top, remains in the funnel and is removed in the same way into another container.

2. The separation of two or more liquids that mix completely is normally carried out by **distillation**. It is based on the idea that the two miscible liquids have different boiling points. The mixture to be separated is placed in a flask attached to a water-cooled condenser. This is a long sloping glass tube surrounded by a water jacket. Cold water from the tap enters at the bottom of the condenser and flows out at the top. The connections are made using rubber tubing. The flask is then heated, usually with a water bath. The more volatile liquid (the one with the lower boiling point), passes off. The liquid with the higher boiling point stays inside the flask.

3. Distillation is also an effective way of purifying a liquid, where dissolved impurities remain in the distillation flask. But this may not be easy if the two boiling points are close together. Also, there are some liquid mixtures that cannot be completely separated by ordinary distillation – a water and ethanol mixture, for example.

Fractional distillation

Fractional distillation can be used to get a nearly pure sample of a liquid where ordinary distillation does not work. In the case of water and ethanol, mentioned above, a nearly pure sample of ethanol can be obtained. The addition to the apparatus used for ordinary distillation is a vertical glass tube packed with glass beads, and a thermometer at the top. When the temperature rises above 78°C, the boiling point of ethanol, the distillation process comes to an end, as only water will remain in the flask.

- *Examples* – Fractional distillation is used on a big scale in the petrochemical industry, to separate the 'fractions' of crude oil. The lower boiling fractions are called octanes. These are produced on a vast scale to be sold as petrol at filling stations. There are also uses for the others. See chapter 9 for more on fractional distillation, and the range of products obtained.

Helping you learn

Progress questions

Summarise the various methods for separating mixtures of substances. How would you try to separate the following mixtures?

1. salt and iron filings (two methods).

2. the oxygen and nitrogen in air.

3. the water in sea water.

Practical assignment

Write a brief descriptive account of the main reasons for the importance of chemistry in our lives. Include as many common substances as you can think of, which are produced in the chemical industry.

2

The Building Blocks of Chemistry

One-minute summary – The building blocks of chemistry are **elements**, **atoms**, **molecules**, **ions** and **electrons**. They are all referred to as 'particles'. They form all matter, whatever its nature. The main ones are atoms, molecules and ions. This topic is very important. You will learn what these terms mean, and the part they play in chemistry. Understanding each one will give you a very good start. Students often confuse these terms, so we begin by making a clear distinction between these, and by understanding the exact meaning of an element. There are some other important particles that make up atoms, but we will look at those later. In this chapter we will explore:

- elements and atoms
- molecules, ions and electrons
- how chemical compounds are formed
- atomic structure
- atomic number, mass number, and the carbon standard
- isotopes
- radioactivity and its uses.

Elements and atoms

Elements are substances that can, and often do, exist as a separate chemical species. Elements are all composed of atoms, or in a few cases molecules containing two atoms. For example, iron, nitrogen and oxygen are all elements.

- *Key point* – The atoms of elements cannot be split up into a simpler substance by chemical means.

► *Key point* – Each element is made up of atoms all of the same type, but different from all other elements.

All other matter is also made up of combinations of these elements. There are over ninety of them. Many of these elements are rare metals.

Compounds and elements

The difference between the elements, and the huge range of other substances referred to as compounds, arises from the following factors:

1. which elements are present in the compound
2. how many atoms of the various elements each one contains
3. how they are joined together.

Symbols of the elements

Each element has its own 'personal' chemical symbol. This helps us identify them. Look at the Periodic Table of the elements on page 122. Examples of these symbols are:

H	hydrogen
S	sulphur
Na	sodium
Cl	chlorine

If there are two letters in the symbol, the first is always a capital, as in the symbol Na for sodium.

Molecules, ions and electrons

Molecules

A molecule is a group of at least two atoms, chemically joined together. Molecules can exist separately. Generally, the substance made up of those molecules will not resemble its parent elements, either physically or chemically.

The simplest molecules

The simplest molecules are those of most gaseous elements:

hydrogen (H) oxygen (O) nitrogen (N)

All of these gases contain two atoms in their molecule, for example H_2. However, all other molecules, in compounds, contain atoms of more than one element. For example, water is H_2O, and it contains both hydrogen and oxygen.

The water molecule is one of the simplest examples of a chemical compound and the way it is written is known as the formula for water.

Many molecules are much larger and more complicated. Vitamin C has the formula $C_6H_8O_6$. As you can tell, it contains three elements and twenty atoms in its molecule. It has six of carbon, eight of hydrogen, and six of oxygen.

Ions

An ion is a charged particle. Note that atoms and molecules are neutral and have no charge. The ion is formed when a particle, usually an atom of an element, loses or gains electrons. The electron carries a single negative charge.

Ions only normally exist either in a solid, or in solution in water *where there are other ions of opposite charge*. An ion can start as a single atom, or group of atoms, which becomes charged by gaining or losing electrons, for example:

$$Na - 1e^- \rightarrow Na^+$$

Cations and anions

Atoms of metals, and hydrogen, *lose* one or more electrons and form positive ions called cations, whilst non-metals *gain* electrons forming negative ions or anions. Groups of negatively charged atoms, or anions, are sometimes also referred to as radicals. They are usually groups of non-metals. Carbonates, hydroxides, nitrates and sulphates are all examples of groups of non-metallic ions. Full details of their charges and formulae can be found in chapter 4.

Ions need partners
It is important to understand that ions cannot normally exist on their own. They must always have, as partners, ions of opposite charge, in the correct numbers to form a neutral substance. Thus salt, sodium chloride, contains two ions of opposite charge, and therefore can exist as such.

How chemical compounds are formed

All chemical compounds are formed from the ninety or so elements, although the 'lighter' elements at the top of the Periodic Table are predominant. So compounds are formed by elements combining with each other.

As mentioned earlier, compounds generally do not resemble the elements they started from and have different physical and chemical properties.

- ► *Example* – Sodium chloride, or common salt, is an excellent example. Sodium is a soft shiny metal and chlorine is a green gas, yet salt is a white solid. The reason is that sodium ions Na^+, and chloride ions Cl^-, are completely unlike their parent atoms.

In the process of ion formation, during a chemical reaction, metals lose electrons and non-metals gain them. Look at these examples:

$$Na - 1e^- \rightarrow Na^+$$

$$Cl + 1e^- \rightarrow Cl^-$$

These are the two ions present in sodium chloride.

Valency

The numbers of electrons lost or gained depends on the elements' combining power, also called **valency**. For elements this is normally the number of electrons its atom must either lose to form a positive ion, or gain to form a negative ion, in order to have a complete outer' shell' of eight electrons. This is known as an 'inert gas configuration'. It is because the inert gases located on the far right of the Periodic Table already have a full outer shell of eight electrons.

This process is central to the formation of all chemical compounds (more in chapter 5). In those compounds the various elements present are bonded together. Mixtures, including alloys of metals, also contain more than one substance but are not chemically bonded together. Generally, although not always, the components of mixtures are fairly easy to separate again.

Atomic structure

All atoms are made up of three basic particles. There are **protons** and **neutrons** in a central nucleus, and **electrons** in relatively remote orbits round the nucleus.

1. The electrons are remote from the nucleus, and contribute little to the mass of an atom. Atoms are largely made up of neutrons and protons.

2. Neutrons and protons are almost identical in mass. They are given a mass of one atomic mass unit. Their actual masses are tiny, so the atomic mass unit is a better way of comparing the masses of different atoms.

Particle charges

The charges on the three particles are as follows:

(a) electrons are negatively charged
(b) protons are positively charged
(c) neutrons are uncharged
(d) the charge on the electron is opposite to, but the same as the charge on the proton.

Figure 1 shows the structure of the sodium atom.

The atomic number of sodium is 11. The nucleus contains 11 protons with 11 electrons, in three orbits around it.

Atomic structure of sodium atom
The atom is uncharged.

Atomic structure of sodium ion
The ion has a single positive charge by loss of one electron from the outermost shell.

KEY
• = electrons
p = protons
n = neutrons

Figure 1. The structure of the sodium atom and sodium ion.

Atomic number, mass number, and the carbon standard

The **atomic number** of an element is equal to the number of electrons, or protons since these are the same. The **atomic mass number** of an atom is equal to the total number of protons and neutrons added together. It is always a whole number.

► *Example* – Sodium contains a total of 11 electrons in three 'shells' around the nucleus. There are 2 in the shell nearest the nucleus, 8 in the next and 1 in the outermost shell. This lone electron gives sodium its considerable reactivity. There are also 11 protons, and 12 neutrons in the nucleus. The atomic number of sodium is therefore 11 and its atomic mass number is 23.

Measuring atomic mass

The atomic mass of all the elements is measured against a chosen standard. This standard is the element carbon. Carbon is given an atomic mass of exactly 12, since its nucleus contains 6 protons and 6 neutrons.

Rather than use actual masses, which are so small, it is much more convenient to compare the atoms of other elements to the chosen standard. The chosen carbon standard is ascribed an atomic mass of 12.0000 units and all other atoms of the elements are compared to this standard. They are therefore called 'relative atomic masses'. More often they are just referred to as 'atomic masses'. A table of the approximate atomic masses of the commoner elements is given at the end of the book.

We can now experimentally evaluate the atomic mass of each element to a very high degree of accuracy.

- *Hydrogen* – occupies a unique position in the Periodic Table. Its atomic number is 1 and contains 1 proton in the nucleus, with a single electron outside it. It is the only element with no neutrons in the nucleus, but see isotopes below.

- *Helium* – the second element, with mass number 4, contains 2 electrons, with 2 protons and 2 neutrons in the nucleus.

Isotopes

Nearly all the elements contain isotopes. Oxygen has 3 isotopes. In the isotopes of oxygen, the number of both electrons and protons remains at 8, in line with its atomic number which is also 8, but there are different numbers of neutrons in the nucleus, as the table shows:

Oxygen isotopes

Mass number	*Number of neutrons*	*% of each isotope*
16	8	over 99%
17	9	0.000 15%
18	10	under 1%

These three isotopes occur naturally in ordinary oxygen, and make its atomic mass not quite a whole number, because of the mixture of isotopes. However, it is normally taken as 16, as this is the predominating isotope. See figure 2 for a diagram of the nuclei of the three isotopes of oxygen.

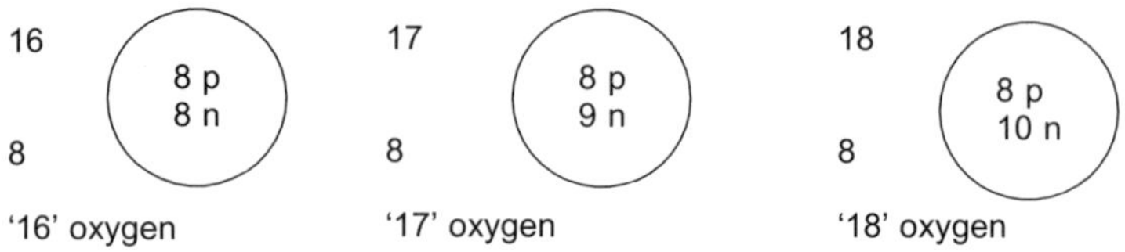

NOTES

1. p = protons, n = neutrons
2. Each isotope contains 8 electrons outside the nucleus.
3. In naturally occuring oxygen (eg. in air), the '16' isotope is largely predominant. Its atomic mass is therefore very close to 16.

Figure 2. The nuclei of the three isotopes of oxygen.

Hydrogen isotopes

Hydrogen also consists of three isotopes. Ordinary hydrogen, with mass number one, contains only one proton in its nucleus. There are two more isotopes containing one and two neutrons respectively, in addition to the proton.

Chlorine isotopes

Chlorine, with atomic number 17, is the best example of all. Remember, each isotope must contain 17 electrons and 17 protons. There are two isotopes only, mass numbers 35 and 37. These therefore contain 18 neutrons (35/17) and 20 neutrons (37/17) respectively. See figure 3 for a diagram showing the chlorine isotopes.

The relative abundance of these two isotopes in naturally occurring chlorine is approximately 75% of the '35' isotope (18 neutrons) and 25% of the '37' isotope (20 neutrons). In naturally-occurring chlorine this leads to a relative atomic mass of about 35.5.

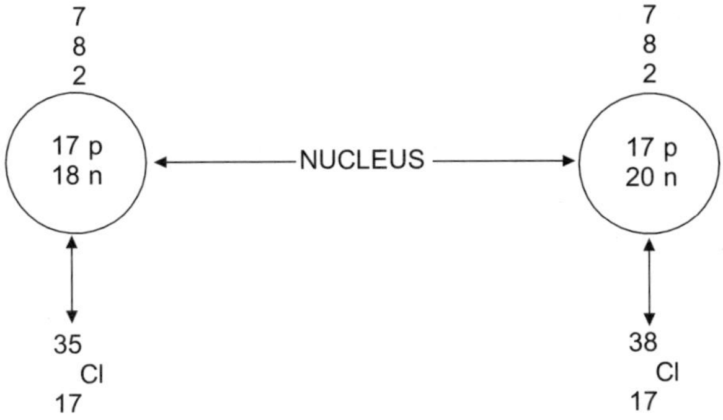

KEY
p = Protons in nucleus.
n = Neutrons in nucleus.
2, 8, 7 = Electronics in 3 'shells' (orbits) around the nucleus.

Figure 3. The two isotopes of chlorine.

Generally, since only the neutrons are involved, the chemical properties of different isotopes of the same element are the same. There is a recognised shorthand way of writing down the configuration of an atom, shown here for the two isotopes of chlorine:

$$^{35}_{17}Cl \text{ and } ^{37}_{17}Cl$$

The top number is the mass number, or the number of protons *and* neutrons added together. The lower number is the atomic number, or the number of electrons or protons.

Radioactivity

Radioactivity refers to the emission of radiation by some elements. Radioactivity may be natural or artificial.

(a) In natural radioactivity some of the heavier elements with very large nuclei are unstable and their nuclei break down. Examples of elements that are naturally radioactive include radium and uranium.

(b) In artificial radioactivity certain elements are bombarded with high-speed particles and become radioactive as a result.

Radioactive particles

There are three types of radioactive particles emitted from the nuclei of radioactive elements. They are called **alpha**, **beta** and **gamma** radiation.

1. Alpha particles are helium nuclei. Loss of this from a nucleus results in the nucleus having an atomic mass number of four less than originally.

2. Beta particles are fast electrons emitted from the **nucleus**, not from the outer electron shells.

3. Gamma rays are radiation of very short wavelength. Radio isotopes are said to decay when they give off radiation. The measure of the speed of decay is the time taken for the measured radiation to have fallen such that half the atoms have decayed.

Uses of radioactivity

Whilst exposure to high level radiation can be very harmful, radio isotopes have a number of valuable uses. For example enriched uranium is being increasingly used in power stations since it produces far more energy per unit of mass than coal. The safe disposal of radioactive waste is, however, a problem.

Other uses are in cancer treatment, also for radioactive tracing and carbon dating of any material containing carbon such as trees and old manuscripts.

Helping you learn

Progress questions

1. Explain the meanings of the following terms: element, atom, ion and molecule.

2. Write down the symbol or formula for each of the following:

   (a) a molecule of bromine
   (b) two atoms of zinc
   (c) a molecule of sulphur dioxide
   (d) a molecule of hydrogen chloride.

3. Make a table showing which of the following form positive ions and which negative ions. Indicate how you arrived at your decision. Potassium, sulphur, copper, hydrogen, barium, bromine, oxygen, and silver.

4. Explain clearly why the inert gases in group 0 of the Periodic Table do not normally enter into chemical combination.

5. Oxygen, atomic number eight, has three isotopes of mass numbers 16, 17 and 18, and its atomic mass is very close to 16. Explain the terms atomic number, isotope, mass number and relative atomic mass, and make diagrams showing its three isotopes.

Study and revision tips

1. Remember that metals (on the left side of the Periodic Table) lose electrons forming positive ions. Non-metals (on the right side of the Periodic Table) gain them forming negative ions.

2. Understand isotopes. Learn how to write down the configuration of an element. Be able to draw diagrams of the atoms of elements up to chlorine. Check from your syllabus how much you need to know about radioactivity.

3. Thorough knowledge of this chapter is essential to your understanding of the rest of the subject. In particular the third practice question will test your grasp of atomic structure. If you went wrong, redo that question.

3

The Periodic Table and Periodicity

One-minute summary – In this chapter you will learn all about the Periodic Table. This is an arrangement of all the elements in order of their ascending atomic number. Remember that **atomic number** is the number of electrons that exist in the atom of any element, outside the nucleus. It also equals the number of protons. We discuss the meaning of Groups and Periods of elements in the Periodic Table and look at the physical and chemical changes that take place in both of these. The Periodic Table is shown on page 122. It shows that, for example, magnesium has the symbol Mg, and it has an atomic number of 12. It has an atomic mass of 24. A table of approximate atomic masses is given at the end of the book. It is well worth getting familiar with the Periodic Table, and knowing the sequence of elements up to about magnesium. Most of the elements, about 80% of the Periodic Table, are metals. Most of the elements on the left side of the Periodic Table are metals and most on the right side are non-metals. Some Periodic Table charts have a 'staircase' shown on them, marking the approximate boundary between metals and non-metals. In this chapter we will explore:

- periodicity: Groups and Periods
- the Groups
- a summary of Groups
- metals and non-metals
- oxidation number
- the reactivity series
- case study: Mendeleev and the Periodic Table

Periodicity: Groups and Periods

It is useful here to have a copy of the Periodic Table to hand. Early chemists noticed that there was a repeating pattern of the properties, both physical and chemical, of the then known elements. After every eight elements, the next one was similar to the first one. This is why the table of the elements came to be known as the Periodic Table.

Terms to understand

1. **Vertical** columns are called **Groups**. Each Group contains elements that have similar physical and chemical properties. This similarity is particularly marked in the groups at the sides of the Periodic Table.
2. **Horizontal** Groups of elements are called **Periods.**
3. Moving from left to right across a period their properties change from metallic to non-metallic in character.
4. Some Groups have names. For example, Group I (far left, Li to Cs) is called the 'alkali metals', and Group VII (F to I) is called 'the halogens'.
5. The large horizontal blocks of metallic elements towards the centre, between Groups II and III are called 'transition elements'.
6. The repeating pattern of properties in a Group is due to the elements in each Group having similar electronic configurations. This means that the arrangement of the electrons in the atom of each is similar, and in particular in the outermost shell.

It is the electrons in the outermost shell which are generally the only ones involved when the element takes part in a chemical reaction.

▶ *Example* – Taking Group I as an example, the elements, all metals, contain a single electron in their outermost shell. The ease with which this single electron is given up accounts for the great reactivity of these elements.

The Groups

Table 1 shows the electronic configuration of the Group I metals. As you can see, each element has a single electron in the outermost orbit, or shell. This electron is very easily given up, forming an ion with a single positive charge.

As the atom increases in size, downwards, so too does its reactivity. Lithium is the least reactive, and caesium the most. This is because the single 'valency' electron is most easily lost where it is furthest away from the nucleus. Comparing the reactions of sodium and potassium towards water soon confirms this. The hydrogen produced by potassium catches fire almost at once, with sodium it usually does not. Table 2 shows some properties of the Group I metals.

Element	*Atomic Number*	*Electronic configuration*
lithium	3	2.1
sodium	11	2.8.1
potassium	19	2.8.8.1
rubidium	37	2.8.18.8.1
caesium	55	2.8.18.18.8.1

Table 1. Configuration of Group I metals.

Name of element	Hardness/melting point	Reactivity
lithium	hardest, highest melting point	brisk with water
sodium	quite soft	violent/rapid
potassium	soft	very violent
rubidium	soft	very wiolent
caesium	very soft, lowest melting point	explodes

Table 2. Properties of Group I metals.

Element	*Atomic number*	*Configuration*
fluorine	9	2.7
chlorine	17	2.8.7
bromine	35	2.8.18.7
iodine	53	2.8.18.18.7

Table 3. Configuration of the Group VII elements.

Halogens

The halogens and their configuration are shown in table 3. The elements of this group, group VII, as they are non-metals, generally combine by accepting an electron from a metal to build up their outermost 'shell' to eight electrons. This takes them to the inert gas configuration. They too show a gradual change in properties as in Group I, but this time the reactivity decreases downwards, so that fluorine is the most reactive and iodine the least. Fluorine is the most reactive of all the non-metals, and possibly of all the elements.

At the end of chapter 6 there is a brief description of the original isolation of fluorine. The halogen ions (see chapter 5) formed in this way have a single negative charge and are called fluorides, chlorides, bromides and iodides. The decreasing reactivity of the halogens can be easily investigated experimentally. There is a practical assignment on this topic at the end of this chapter.

Groups II to VI

Similar comments apply to Groups II to VI. However the elements of Group IV, headed by carbon, generally do not form ions, instead preferring to share their electrons in the process of compound formation. This is discussed in chapters 5 and 6.

Group 0

The last group of elements, group 0 on the far right of the Periodic Table, is the inert gases. They already have full outer orbits and thus do not enter into normal chemical combination.

A summary of Groups

1. It is useful to remember that the Group Number is also equal to the number of electrons in the outer, or 'valency' orbit of the elements in that group. All the ions of the elements in a group have the same charge.

2. Also remember that elements combine to try and get to a configuration which has eight electrons in the outermost shell, either by losing or gaining electrons.

3. Look carefully at table 4. This shows the bonding and valency (combining power) of elements in the groups. Those groups at the extremities of the Periodic Table form ionic bonds. Those in the centre tend to form covalent bonds.

Group Number	*Type of bond formed*	*Normal valency*
I	Ionic	1
II	Ionic	2
III	Generally ionic	3
IV	Generally covalent (shared electrons)	4
V	Both types	3
VI	Generally ionic	2
VII	Generally ionic	1
0	No bonds	0

Table 4. Bonding and valency for the Groups.

Metals and non-metals

There are a number of charts of the Periodic Table, showing varying amounts of information. Some types have a 'staircase' towards the right hand side of the Table. This is meant to show a rough 'dividing line' between metals and non-metals. Elements to the left of the staircase are metals, and those to the right are non-metals.

However, some of the elements next to the staircase show

properties of both types. The element arsenic is a good example. Note its position in the Periodic Table, with the symbol *As*.

We have mentioned metals and non-metals several times. Table 5 compares the physical and chemical properties of metals and non-metals.

	Physical properties	
Property	*Metals*	*Non-metals*
melting point	usually high	usually low
hardness	usually hard	usually brittle
tensile strength	usually strong	low, break easily
electrical conduction	usually good	poor, or non conductors
malleability	good	very low
ductility	good	low, fracture easily
	Chemical properties	
nature of oxide	solids, basic (i.e. they neutralise acids)	gases, or low melting point solids – usually give acidic solutions with water
with dilute acid	often give hydrogen	no reaction
during electrolysis	ions go to cathode	ions go to anode

Table 5. The properties of metals and non-metals.

Malleability means it can be beaten into a sheet. Ductility means it can be drawn into a wire.

Oxidation number

Check with your teacher, or a copy of your syllabus, to find out whether 'oxidation number' is required by your examination board.

Definition

'Oxidation number' means the number of charges (positive, negative or zero) on an atom, an ion, or the elements in a compound. For atoms of the elements it is zero, as they carry no charge. For the ions of the elements it is the same as the valency or combining power.

If the formula for a substance is known, and the oxidation numbers of all the elements except one is also known, we can work out the oxidation number of the remaining element. Some simple examples will show this. However, it means that you need to learn the formulae for all the common ions.

Example

To calculate the oxidation number of sulphur in the compound potassium sulphate:

1. Write down the formula:

$$K_2SO_4$$

2. Write down the values of the oxidation numbers of the other elements. In this case:

$$K = +1, \text{ and } O = -2$$

3. Write down $K_2SO_4 = 0$. (It is always zero for a complete formula.)

4. Then substitute the values of the oxidation numbers in the formula:

$$2 \times (+1) + S + 4 \times (-2) = 0$$
$$2 + S - 8 = 0$$

from which $S = +6$

It can be arrived at more simply thus:

$$SO_4^{--} = -2$$

Remembering that SO_4^{--} is the formula for the sulphate ion, then:

$$S + 4 \times (-2) = -2$$
$$S - 8 = -2$$
$$\text{ie } S = +6$$

Another worked example

What is the oxidation number of chromium in the substance potassium dichromate?

The formula is $K_2Cr_2O_7$, thus $K_2Cr_2O_7 = 0$

The other Oxidation numbers are: $K = +1$, $O = -2$

$K_2Cr_2O_7 = 0$ as set out above

Thus $2(+1) + 2(Cr) + 7(-2) = 0$

$\therefore 2Cr - 12 = 0$, from which $Cr = +6$

Therefore the oxidation number of chromium in potassium dichromate is + 6.

The reactivity series

The reactivity series is a list, in order of descending reactivity, of the most common metals. The list is arranged according to their reactivity towards oxygen, water and dilute acids. It is shown in table 6.

Metal	*Symbol*	*In oxygen*	*Cold water*	*Dilute acid*
potassium	K	burns easily	very rapid	explodes
sodium	Na	burns easily	very rapid	explodes
calcium	Ca	burns	brisk	very rapid
magnesium	Mg	burns	slow	very brisk
aluminium	Al	burns as powder	very slow	very brisk
zinc	Zn	burns as powder	very slow	quite brisk
iron	Fe	slow, rusts	burns in steam	quite brisk
(hydrogen)	H			
copper	Cu	very slow	hardly any	very little
mercury	Hg	very little	none	hardly any
gold	Au	none	none	none

Table 6. The reactivity series.

Notes:

1. The metals at the top (K, Na and Ca) react with cold water. Potassium is above sodium because the reaction is quicker, and the hydrogen produced catches fire. This does not normally happen with sodium.

2. The metals below calcium react with either hot water or steam. In all cases hydrogen is produced, plus the hydroxide or oxide of the metal. Those below hydrogen do not displace it from water or dilute acids.

3. The metals at the top burn in either air or oxygen to form strongly basic oxides. Those at the bottom only oxidise slowly in oxygen and do not burn.

4. The metals at the top are the strongest reducing agents. They will displace the metals below them from solutions of their ions. For example, zinc displaces copper from its ions:

   zinc + copper ions = zinc ions + copper

   The copper ions can be from a solution in water of any copper compound. The original blue colour of the solution disappears as the copper ions are removed in the reaction.

5. Note that even the metals below hydrogen in the reactivity series dissolve in nitric acid, because this acid is an oxidising agent. Hydrogen is not normally produced in the reaction between nitric acid and metals.

6. The reactivity series explains why the metals above hydrogen do not occur as the free elements in the ground, and why those metals at the top have to be extracted by electrolysis. Electrolysis is discussed in chapter 6. Those metals close to hydrogen, such as zinc and iron, are normally extracted by reduction of the oxide with carbon in a furnace. Iron is prevented from rusting by the application of a layer of zinc. This process is called **galvanising**. The zinc reacts with water vapour and oxygen in air, thus preventing rusting. However, if the protective layer of zinc is broken, rusting proceeds even faster.

Case study: Mendeleev and the Periodic Table

A number of chemists from the nineteenth century had noticed the repeating properties of the known elements. Mendeleev, who was Russian, tried to construct one of the early forms of the Periodic Table. He even left spaces in his table for hitherto undiscovered elements. He made an accurate forecast in some cases of the likely properties, both physical and chemical, of the missing elements. For example, he left a gap below silicon (Group IV) for an unknown element. His predictions were very close when the element germanium was eventually discovered.

Helping you learn

Progress questions

1. Imagine that you have samples of arsenic chloride $AsCl_3$ and arsenic oxide As_20_3. Describe a series of tests for both substances, which you could carry out to decide whether arsenic is a metallic or non-metallic element.

Discussion points

1. What can you tell about the element strontium and its compounds, from its position in the Periodic table in Group II? Consider the reactivity of the element, and the nature of its compounds, in terms of bonding.

2. Can you predict the physical and chemical properties of the element astatine, which is located at the bottom of Group VII?

Practical assignment

An investigation of the displacement of one halogen by another

Prepare solutions in water of sodium bromide and sodium iodide. Place in two test tubes. Make a small quantity of chlorine gas by warming concentrated hydrochloric acid and manganese dioxide in a test tube.

Pass the gas so generated (taking care, because chlorine is toxic) into the two prepared solutions. The bromide ion solution will turn red with the formation of free bromine and the iodide ion solution will turn brown with the formation of free iodine. This establishes that chlorine is more reactive than either bromine or iodine as it has displaced them. Finally, obtain some bromine water (a small quantity, about one drop – of bromine added to 3 cm of water in a test tube) and add the solution to the iodide solution.

Observe the results. Why do you think the use of fluorine has not been mentioned?

Study and revision tips

1. Make sure you understand the construction of the Periodic Table. The metals are mainly on the left, and the non-metals on the right.

2. Elements enter into chemical combination by losing electrons (metals) or gaining electrons (non-metals). They do this to finish up with a complete outermost shell of eight electrons. The number of outer electrons is also the same as the Group Number. The elements are arranged into vertical groups and horizontal periods. The elements in a particular group are generally similar in both physical and chemical properties, but show a gradual change down the Group. In the case of metals reactivity increases downwards, with the non-metallic Groups it decreases downwards. Across a period (left to right) there is a change from metals to non-metals.

3. This is a very popular topic with examiners, so make sure you are familiar with both your syllabus content, and the content of this chapter. You should be able to predict the properties of an unknown element from the Periodic Table.

4. Make sure you understand the reactivity series, and know the positions of the common metals in it. Hydrogen is normally included since, like metals, it forms positive ions.

4

Chemical Formulae and Equations

One-minute summary – This is an important chapter. Here you will learn all about chemical formulae. You will discover how to write down formulae, and also how to write chemical equations. You will find out what information is revealed by an equation. The chapter explores the following topics:

- chemical formulae
- equations and how to balance them
- the mole and molar mass
- calculating quantities from equations
- the reverse calculation
- the molar volume of gases
- calculations involving solutions, usually acids and alkalis.

Chemical formulae

You need to be able to write down the formulae of common chemical compounds, especially ionic compounds. Table 7 shows the formulae of the commonest ions, and the charge they carry. It will pay you to learn this table off by heart. The pattern is easy to follow for metallic ions from their position in the Periodic Table.

Once you have learned it, writing down the complete formula for an ionic compound is simple. Just write down the formulae of the two ions, and then add suitable numbers to *make the charges balance*. Here are some examples:

1. Sodium chloride. The ions are Na^+ and Cl^-, and the formula is NaCl.
2. Potassium sulphate. The ions are K^+ and SO_4^{--} and so the formula is K_2SO_4.

Cations (metallic ions)		*Anions (non-metallic ions)*	
K^+	potassium	Cl^-	chloride
Na^+	sodium	Br^-	bromide
Ca^{++}	calcium	I^-	iodide
Mg^{++}	magnesium	OH^-	hydroxide
Al^{+++}	aluminium	CO_3^{--}	carbonate
Zn^{++}	zinc	HCO_3^-	bicarbonate
Fe^{++}	iron II	SO_3^{--}	sulphite
Fe^{+++}	iron III	SO_4^{--}	sulphate
H^+	hydrogen	HSO_4^-	bisulphate
Pb^{++}	lead	NO_3^-	nitrate
Cu^{++}	copper	O^{--}	oxide
Ag^+	silver		
NH_4^+	ammonium		

Table 7. Ions and their charge.

3. Magnesium hydroxide. The ions are Mg^{++} and OH^- and the formula is $Mg(OH)_2$. In this case note the bracket, indicating that there are two $(OH)^-$ ions in the compound, that is, a total of five atoms. $MgOH_2$ is not correct.

Equations and how to balance them

Let's start with a word equation. This shows in words the starting substances on the left, an arrow showing the direction of the chemical reaction, and the products of the reaction on the right:

nitrogen + hydrogen → ammonia

Sometimes an = sign is inserted between reactants and products. An arrow has the advantage of showing the direction of the reaction. To turn this into a chemical equation we need to know the formulae of the three substances involved. These are nitrogen

(N_2), hydrogen (H_2), and ammonia (NH_3). Now we can write an equation:

$$N_2 + H_2 \rightarrow NH_3$$

But there are *two* nitrogen atoms on the left, and only one on the right. So we must multiply the NH_3 by two to get two nitrogen atoms on the right.

$$N_2 + H_2 \rightarrow 2NH_3$$

Notice that there 6 hydrogen atoms on the right, so finally multiply the hydrogen on the left by 3 to get the correct equation:

$$N_2 + 3H_2 \rightarrow 2NH_3$$

► *Key point* – You can only insert multiplying numbers at *the front of the formula for each substance.* You cannot change the formula of any substance in the equation. This process, which is very common and important, is known as **balancing**, and it is a vital first step in making calculations from equations.

Practice exercise 1

Here is an exercise to try. See if you can balance this equation:

$$C_2H_2 + O_2 \rightarrow CO_2 + H_2O$$

It has become customary, in writing equations, to add 'state' symbols to show whether the substances in the equation are solids, liquids, or gases. The symbols are (s) for a solid, (l) for a liquid and (g) for a gas. The symbol (aq) is often used for a substance dissolved in water, for example, to distinguish between a dilute and a concentrated acid.

What the equation tells us

In our equation above, showing the formation of ammonia, all three substances are gases, so it is written:

$$N_2\ (g) + 3H_2\ (g) \rightarrow 2NH_3\ (g)$$

The arrow tells us the direction of the change, from left to right.

In addition the equation tells us another important piece of information. Assuming that the reaction has taken place completely, 1 mole of nitrogen (28 grams) reacts with 3 moles of hydrogen (6 grams) to produce 2 moles (34 grams) of ammonia. Thus the masses on left and right hand sides should always be the same, as long as the equation is properly balanced.

However, the equation does *not* tell us whether it is complete. We do not know whether all the nitrogen and hydrogen been used up. Nor do we know what external conditions are needed to form ammonia, or whether heat is given out or absorbed in the reaction.

Here is another example:

$$2Na\ (s) + 2H_2O\ (l) \rightarrow 2\ NaOH\ (aq) + H_2\ (g)$$

As discussed above, no other information is given. In fact, this particular reaction takes place at room temperature, that is, with cold water.

Ionic equations

Ionic equations are limited to those reactions that take place in **aqueous solution.** The reactants and products are written out as their ions in the aqueous solution, and the 'passenger ions' are then eliminated as they are there before and after the reaction.

Ionic equations are dealt with in detail in chapter 7.

The mole and molar mass

The purpose of the mole is to compare equal numbers of atoms, since equal masses of two different substances is not a useful comparison to chemists.

Definition of the mole

The 'mole' is simply the formula mass expressed in grams. It is also called molar mass and formula weight. Here are some simple examples:

Substance	*Formula*	*Molar mass*
water	H_2O	18g
sulphuric acid	H_2SO_4	98g
magnesium bromide	$MgBr_2$	184g

This quantity is obtained by simply adding together the atomic masses of all the atoms in the molecule. The atomic masses are, for sulphuric acid, H = 1, S = 32 and O = 16. So the mass of one mole of sulphuric acid is (2 x 1) + 32 + (4 x 16) = 2 + 32 + 64 = 98g.

Here are some more examples of common substances.

Substance	*Formula*	*Molar mass*
sodium chloride	NaCl	23 + 35.5 = 58.5g
potassium nitrate	KNO_3	39 + 14 + (3x16) = 101g
calcium carbonate	$CaCO_3$	40 + 12 + (3x16) = 100g
copper sulphate	$CuSO_4.5H_2O$	63.5 + 32 + (4x16) + (5x18) = 249.5g
lead hydroxide	$Pb(OH)_2$	208 + 2(16+1) = 242g

Calculating quantities from equations

The following examples show how to use an equation to calculate quantities. Remember, the starting point is always a correctly *balanced equation.* (Atomic masses are required. They can be found in the table at the end of the book.)

Example 1

What mass of lead oxide and carbon dioxide are obtained by heating 20g of lead carbonate $PbCO_3$?

$$PbCO_3\ (s) \rightarrow PbO\ (s) + CO_2\ (g)$$

$$(208g + 12g + 48g) = (208g + 16g) + (12g + 32g)$$

$$268\ g = 224\ g + 44\ g$$

$$1g = \frac{224g}{268} + \frac{44g}{268}$$

$$20g = \frac{224 \times 20}{268g} + \frac{44 \times 20}{268g}$$

$$20g = 16.72\text{ g and } 3.28g$$

Therefore 20g of lead carbonate gave 16.72g of lead oxide and 3.28 g of carbon dioxide. It is important to note that if the totals on the two sides of the equation are not the same, then you have made an error.

Example 2

In the neutralisation reaction as shown:

$$NaOH\ (aq) + HCl\ (aq) \rightarrow NaCl\ (aq) + H_2O\ (l)$$

20g of sodium hydroxide was mixed with 20g of hydrochloric acid. One of these was in excess (i.e. some of it was left over). Which one was in excess and how much was left over?

Start by writing out the equation:

NaOH (aq)	+	HCl (aq)	→	NaCl (aq)	+	H_2O (l)
1 mole		1 mole		1 mole		1 mole
40 g		36.5 g		58.5 g		18 g

Quickly check that masses on both sides are equal.

Therefore 20 g NaOH reacts with $\frac{36.5}{2g}$ = 18.25g of HCl.

Therefore 18.25 g of HCl was needed.

Since we started with 20g of HCl, the HCl was in excess. The amount left over was (2018.25)g = 1.75g. The same basic approach applies to all examples of this type of calculation.

Calculation of simplest formula

If we know the percentage composition of a substance, we can easily work out its formula. The simplest formula shows the ratio of the numbers of the elements present. The true (molecular) formula may be a multiple of the simplest formula.

Example 3

The percentage composition of sulphuric acid is as follows:

hydrogen 2.05%, sulphur 32.65%, oxygen 65.3%. What is its formula? The steps are as follows:

Step one – Divide each % by the relative atomic mass (RAM) of the element:

hydrogen	sulphur	oxygen
2.05%	32.65%	65.3%
RAM = 1	RAM = 32	RAM = 16
$\frac{2.05}{1}$	$\frac{32.65}{32}$	$\frac{65.3}{16}$
2.05	1.025	4.10

Step two – Divide the result for each element by the lowest, ie:

2.05/1.025 = 2   1.025/1.025 = 1   4.1/1.025 = 4

Step three – Write down the formula from these values. It is H_2SO_4

Percentage composition and the reverse calculation

The reverse calculation is sometimes required, that is, converting a known formula and given relative Atomic masses into a percentage composition. Here is an example:

Example 4

What is the % composition of hydrogen, nitrogen and oxygen in nitric acid? Here are the steps in this calculation:

1. Write down the formula of the compound, HNO_3
2. Calculate its formula mass
3. Calculate the % of each element as a % of the formula mass.

Step 1 – formula is HNO_3
Step 2 – formula mass is 1 + 14 + (3 x 16) = 63g

Step 3 – the percentage of each element in the compound is:

Hydrogen	Nitrogen	Oxygen
1/63 x 100	14/63 x 100	48/63 x 100
= 0.0159 x 100	= .2222 x 100	= .7619 x 100
= 1.59%	= 22.22%	= 76.19%

The total of these should of course add up to 100%. If not, then you have made an error!

Example 5

Experimental determination of the formula of magnesium oxide

The experimental method is to carefully burn magnesium ribbon to magnesium oxide. It is accurately weighed beforehand. The ribbon is burned in a closed container to prevent loss of 'smoke' during combustion. The matter is weighed at the end, and the formula can then be worked out as before. Here are some specimen results:

Mass of magnesium started with = 1.2g
Mass of magnesium oxide at the end = 2.0g
Therefore mass of oxygen = 2.0g –1.2g = 0.8g
Thus 1.2g magnesium combined with 0.8g oxygen
Convert to moles of each using Mg = 24 and 0 = 16
1.2/24 moles of magnesium and 0.8/16 moles of oxygen
0.05 moles magnesium combined with 0.05 mole oxygen
Thus 1 mole magnesium combined with 1 mole of oxygen
And the formula is therefore MgO.

Example 6

Ammonium sulphate $(NH_4)_2SO_4$ and ammonium nitrate NH_4NO_3 can both be used as nitrogenous fertilisers. Which is the best one in theory?

As both are easily water soluble, we just need to calculate the nitrogen content in both, to discover which has the higher % of nitrogen. This is done in exactly the same way as for nitric acid earlier on.

Step 1 – Write down both formulae

Ammonium sulphate is $(NH_4)_2SO_4$ and Ammonium nitrate is NH_4NO_3

Step 2 – Calculate the formula mass of both compounds

ammonium sulphate	ammonium nitrate
$2(14+4)+32+(4x16) = 132g$	$14 + 4 + 14 + (3x16) = 80g$
% of nitrogen = 28/132 x 100	% of nitrogen = 28/80 x 100
= 21.2%	= 35%

Thus ammonium nitrate contains a considerably higher percentage of 'fixed' nitrogen (nitrogen in a compound) than ammonium sulphate and is theoretically a better fertiliser.

▶ *Think* – There are two reasons why it is not normally used. What do you think these reasons might be?

The molar volume of gases

Many chemical changes result in gas production. Their lightness makes accurate weighing difficult, but masses of gases can be worked out in an easy way, simply by measuring the volume of gas produced.

A standard of measurement

This volume varies with both temperature and pressure and so a standard is used. The standard is normally:

▶ 1 atmosphere pressure (76cm in a mercury barometer) and room temperature (25°C).

These are abbreviated to stp (standard temperature and pressure). It is also referred to as room temperature and pressure (rtp). A gas volume collected at some other temperature and pressure will need correction, to the volume it occupies at stp.

▶ *Key point* – It is found that, under these standard conditions, *1 mole of any gas fills the same volume.*

This is found to be about 24 dm^3 (litres). So 1 mole of hydrogen (2g), 1 mole of oxygen (32g), or 1 mole of sulphur dioxide (64g) all occupy about 24 dm^3 at stp. Therefore, by measuring the volume of a gas it can easily be converted to a mass, or moles, of any gas generated in a reaction. Below is a worked example:

Example 7
4.8g of magnesium ribbon was completely dissolved in dilute hydrochloric acid. What volume of hydrogen gas (at stp) was produced?

As usual, the first step is a balanced equation:

Mg (s) + 2 HCl (aq) $MgCl_2$ → (aq) +	H_2 (g)
1 mole Mg	1 mole H_2 gas
24g	24 dm^3 H_2 gas at stp
1g Mg	$\frac{24\ dm^3}{24} = 1\ dm^3$
4.8g Mg	$1 \times 4.8 = 4.8\ dm^3$

It should be noted that the *mass* of hydrogen formed is directly obtainable from line 2 in the above calculation.

Measuring the volume of gas

Gas volumes are measured in a number of ways. One example is the Hoffmann water voltameter (See chapter 6 on electrolysis). Another is the gas syringe, which gives reasonably accurate results for small volumes.

Calculations involving solutions

In calculations involving solutions, there is a special – but very convenient – way of expressing the concentration of a solution. This is its **molarity**.

► *Key point* – Any solution which is exactly molar in strength contains exactly one mole (in g) of the solute in 1 dm^3 (litre) of solution.

To work out the number of moles in any solution, this expression is used:

$$\text{number of moles} = \text{volume } (cm^3) \times \frac{\text{molarity}}{1000}$$

Titration

The process of titration is known as volumetric analysis. The idea is to determine, experimentally, the concentration (molarity) of a solution, such as an acid or an alkali. We can then use this to work out the molar mass of the solute in the unknown solution, from the definition of a molar solution given above. A solution of accurately known concentration is known as 'a standard solution'.

Experimental details

In an acid-alkali titration, an accurate volume (usually 25 cm^3) of the alkali is placed in a conical flask (from its shape) with a pipette. Make sure you are familiar with both pipettes and burettes. The solution to be titrated with this is placed in a burette, which can deliver an accurate volume of liquid. Some indicator, usually only two or three drops, is placed in the flask. Neutrality between the acid and alkali is reached when the indicator changes colour.

The volume of acid delivered from the burette (where the acid is usually placed) is then carefully read from the markings on its side. Some skill is needed to obtain accurate results.

A specimen calculation will show how the results are used. Lets say you wish to determine the molarity of some dilute sulphuric acid. In the experiment, 25 cm^3 of sodium hydroxide of molarity 0.12 M was titrated with the dilute acid, and 26.5 cm^3 was required for neutrality.

Step 1 – Write a *balanced* equation:

$$2NaOH\ (aq) + H_2SO_4\ (aq) \rightarrow Na_2SO_4\ (aq) + 2H_2O\ (l)$$

Step 2 – Write down the numbers of moles of each:

2 of sodium hydroxide and 1 of sulphuric acid, from the above equation.

Step 3 – Ratio of moles

$$\text{ratio of moles} = \frac{2}{1} = \text{vol x } \frac{\text{molarity}}{1000}\ (\text{NaOH})$$

divided by vol x molarity (H_2SO_4)

Therefore $\frac{2}{1} = \frac{25 \times 0.12 \times m}{26.5}$ (m = unknown molarity)

$2 \times 26.5 \times m = 25 \times 0.12$

$m = \frac{25 \times 0.12}{2} \times 26.5 = \frac{3.0}{53} = 0.057$ M

Therefore the molarity of the sulphuric acid was 0.057.

Helping you learn

Progress questions

Note: Obtain relative atomic masses from tables.

1. Balance the following equation:

$$NH_3\ (g) + O_2\ (g) \rightarrow NO\ (g) + H_2O\ (g)$$

2. Calculate the molar mass of lead nitrate $Pb(NO_3)_2$

3. Calculate the molar mass of hydrated copper sulphate, $CuSO_4.5H_20$

4. Calculate the percentage composition of all four elements in sodium bicarbonate, $NaHCO_3$.

5. What volume of hydrogen gas was generated (at stp) when 5g of zinc was dissolved completely in dilute hydrochloric acid? (Zn = 65, gram molar volume at stp = 22.4 dm^3).

6. 25 cm^3 of 0.095 M sodium hydroxide was titrated with 27.5 cm^3 of nitric acid at neutrality. What was the molarity of the nitric acid?

7. Here is a word equation showing the action of heat on lead nitrate:

Lead nitrate gives lead oxide, nitrogen dioxide and oxygen.

Convert it into a balanced equation, including state symbols.

Study and revision tips

1. Make sure you understand the definition of the mole and a molar solution.

2. Make sure you can:

   (a) calculate molar mass. (Atomic masses are always given.)

   (b) balance equations. This, (and writing down a balanced equation) is the vital first step in most calculations.

   (c) do calculations involving gas volumes.

   (d) do calculations involving volumes of liquids.

3. All of the topics covered in this chapter are popular with examiners. The secret is plenty of practice, as with anything you want to do well. Remember the importance of being able to write balanced equations. Go back to the one in the text if you were unsuccessful the first time.

5

Bonding of Elements and Compounds

One-minute summary – In this chapter you will learn about the 'links', called **bonds**, which bind together atoms in the elements and compounds in chemistry. At the end you will find a short section on energy in chemistry, since there is a connection between energy and bonding. Remember, in any chemical change existing bonds are broken and new ones formed. In this chapter we will cover:

- electrovalency and covalency
- the general properties of ionic compounds (salts, acids, and bases)
- the general properties of covalent compounds (most other compounds)
- energy changes in chemistry.

Electrovalency and covalency

Chemists distinguish two main types of bond between the atoms in compounds. These are referred to as electrovalent and covalent bonds and the compounds that contain these bonds as electrovalent and covalent compounds. Electrovalent bonds are also referred to as ionic bonds – these terms have the same meaning. There are other types of bonding, for example that found in metals.

The nature of ionic (electrovalent) and covalent bonds

Electrovalent bonding

In ionic compounds, (that is, salts both as a solid and in aqueous solution and also dilute acids, and alkalis), electrons have been completely transferred from a metal to either a non-metal or a non-metallic radical. In the process the non-metal, or the non-metallic radical becomes negatively charged. This process normally takes place when the **metals** on the left side of the Periodic Table combine with the **non-metals** on the right.

Covalent bonding
Non-metals also combine with each other. For example, phosphorus from Group V combines with chlorine from Group VII to form two chlorides, but the bonds are not ionic. They are instead **covalent** bonds, where one or more electrons from each atom form a shared electron pair situated between the nuclei of the two atoms at either end of the bond.

Other bonding
However, in the nature of their bonding, a great many compounds fall somewhere between the two types. In other words, they contain bonds that show the characteristics of both types.

The most 'perfect' ionic bonds are those formed between Group I metals and Group VII the halogens. The most perfect covalent bonds are in the molecules of the gaseous elements, as in the hydrogen molecule H_2, where the shared electron pair sits midway between the two nuclei.

Figure 4 shows what happens when sodium and chlorine combine to form sodium chloride. This is the general pattern in the formation of all ionic compounds, where, as mentioned earlier, both metal and non-metal attempt to gain an outer 'shell' of eight electrons. The eventual charge on each ion depends on how many electrons have been lost or gained.

The general properties of ionic compounds

Ionic compounds
Ionic compounds are salts, acids, and bases.

Structure
The structure is crystalline, with a lattice of ions extending throughout the crystal.

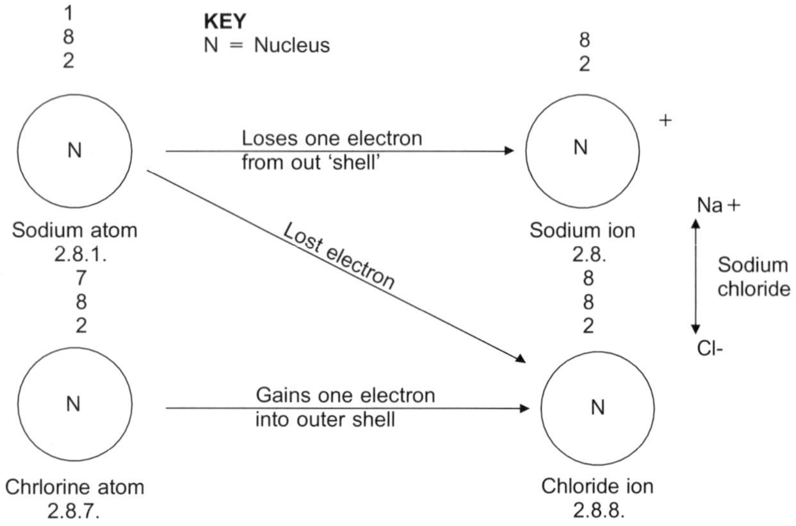

SUMMARY

$Na - 1e^{-} \longrightarrow Na^{+}$

$Cl + 1e^{-} \longrightarrow Cl^{-}$

Figure 4. The changes which take place when sodium and chlorine combine to form sodium chloride.

Melting points

The melting points are generally high, since ionic bonding is strong.

Solubility in water

Most ionic compounds dissolve in water. There is only a low solubility in organic solvents such as ethanol.

Conductivity of solution in water

The aqueous solution is a good conductor. The molten ionic compound is also a good conductor.

General properties of covalent compounds

Covalent compounds are most other compounds, aside from salts, acids, and bases.

Structure
Covalent compounds are generally single molecules. The intermolecular forces are weak.

Melting points
Melting points are usually low. They are very often liquids or gases at room temperature.

Solubility in water
They are often not soluble, but usually dissolve in organic solvents, such as ethanol or benzene.

Conductivity of solution in water
Covalent compounds are poor conductors, or non-electrolytes in solution in water or when molten, since there are no ions.

Energy changes in chemistry

A fundamental principle of chemistry is that, in any chemical change, existing bonds are broken and new ones formed. When this takes place there is an accompanying energy change.

Chemists distinguish two types of reaction, exothermic and endothermic:

- *In exothermic reactions* – heat is given out (from left to right in the reaction equation). New bonds are formed, and the products have less energy than the starting substances.

- *In endothermic reactions* – heat is absorbed (from left to right in the reaction equation). The products have more energy than the starting substances.

All combustions are therefore exothermic. For example:

$$2H_2\ (g) + O_2\ (g) \rightarrow 2H_2O\ (g)$$

A lot of energy is given out in this chemical change.

This next reaction, however is very endothermic:

$$N_2\ (g) + O_2\ (g) \rightarrow 2NO\ (g)$$

A high temperature is needed to make the left-to-right reaction take place. This reaction was once used in the manufacture of nitric acid, although now it is made from ammonia. Figure 5 shows these two processes as energy level diagrams.

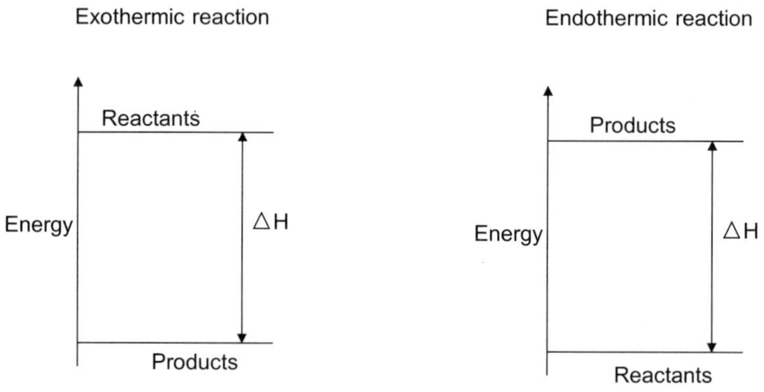

Δ H = Energy difference

Reactants have more energy than products - heat is given out

Ex. $C(s) + O_2(g) \rightarrow CO_2(g)$

= Energy difference

Products have more energy than reactants – heat is absorbed

Ex. $N_2(g) + O_2(g) \rightarrow 2NO(g)$

Figure 5. Energy level diagrams for exo- and endothermic reactions.

Many exothermic reactions, including the burning of fuels, need to be kickstarted. In other words, they do not start until energy is initially supplied. Once started the energy produced keeps the reaction going.

Other energy changes

Most reactions in aqueous solution involve energy changes, but these are often quite small. The precipitation of insoluble salts is an example of a reaction with a small energy change.

One important reaction in which there is a significant exothermic change is neutralisation. This takes place when acids and alkalis are mixed:

$$HNO_3\ (aq) + NaOH\ (aq) \rightarrow NaNO_3\ (aq) + H_2O\ (l)$$

If the 'passenger' ions, that is Na^+ and NO_3^- are removed this reaction becomes:

$$H^+\ (aq) + OH^-\ (aq) \rightarrow H_2O\ (l)$$

The amount of heat energy produced, per mole of water formed, is almost identical for all strong, fully ionised, acids and alkalis. This supports the theory that they are fully ionised in dilute aqueous solution, and that the H^+ ion is present in all acids and the $0H^-$ ion is in all alkalis.

Helping you learn

Progress questions

1. What type of bonding exists in the following compounds: lead bromide, sodium iodide, methanol, sodium hydroxide, glucose, carbon disulphide, hydrogen chloride gas dissolved in the organic solvent toluene, naphthalene and ammonium iodide.
2. Describe experiments you could do to confirm your ideas on the bonding contained in the above compounds.
3. Explain the part that water plays when the gas hydrogen chloride is dissolved in water.
4. What heat change will take place when aqueous sodium hydroxide is mixed with ethanoic acid, a weak organic acid?

Key points to understand

1. The differences between ionic and covalent bonding is that in the former electron transfer occurs from metal and non-metal; in covalent bonding the bond consists of a shared electron pair, generally one from each atom.
2. When metals and non-metals combine together the bonding is nearly always ionic. Dilute acids, bases and salts are ionic

compounds. When non-metals combine together the bonding is covalent.

3. Ionic compounds have strong ionic forces binding them into a large three-dimensional lattice. Their melting points are usually fairly high. Covalent compounds contain weakly bonded individual molecules. Their melting points and boiling points are generally low.

4. In exothermic and endothermic reactions, heat is given out, or absorbed. In neutralisations the heat generated, for strong (fully ionised) acids and alkalis is nearly the same.

Practical assignments

Test the conductivity of 0.5M hydrochloric acid (ordinary bench acid diluted by a factor of 4.) The circuit used for this experiment can be found in chapter 6.

1. Mix together roughly equal volumes of bench dilute HCl and Na0H (about 1 cm^3 of each) in a test tube. Measure the temperature before and afterwards.

2. Make a solution of the gas hydrogen chloride in the organic solvent, toluene. Test the conductivity as in assignment 1 above. (Hydrogen chloride gas is easily generated from solid sodium chloride and concentrated sulphuric acid; seek advice from your teacher to assist you to carry out this experiment).

Study and revision tips

Chemical bonding is another important topic, and popular with examiners.

1. Make sure you clearly understand the differences between ionic and covalent bonding, the elements that form them, and be able to predict the type of bonds likely to be present in compounds from the positions of their component elements in the Periodic Table.

2. In addition, make sure you know the major differences in properties between compounds containing each type of bond.

6

Electrolysis: Splitting Up

One-minute summary – Electrolysis is the term used to describe the chemical changes that result from the effect of electricity on substances. In these chemical changes new substances are formed. Electrolysis takes place when an electric current passes through conducting substances, normally when they are molten, or in a solution in water. The results are usually chemical changes in the substance through which the electric current is passing. These changes would not take place without the electric current. Electrolysis has a number of important industrial applications. In this chapter we will explore:

- terms used – definitions of electrolysis, electrolyte, electrodes
- industrial applications of electrolysis.
- oxidation and reduction in electrolysis
- the 'water' voltameter
- case study: the persistent chemist.

Terms used

Electrolysis

Electrolysis is the decomposition of a compound by passing an electric current (direct, not alternating) through it. The compound must be in fused (molten) form, or be a solution in water. The ions (see chapter 2) in the liquid carry the current.

Electrolyte

Electrolyte is the name of the conducting solution which is split up. Most acids, bases and salts, which are made up of ions (see chapter 2), are electrolytes. Some electrolytes are better conductors than others. Non-conducting liquids and solutions are called non-electrolytes.

Electrodes

Electrodes are the points where the current, from the power source, enters and leaves the electrolyte. The diagram below shows the basic electrolysis circuit. A switch, and bulb to show current flow, may be added. The electrodes must be made of conducting material, commonly some metals and graphite. Normally, they should not be affected chemically by the electrolyte. The connection to the negative terminal of the power supply is the cathode, and to the positive terminal, the anode.

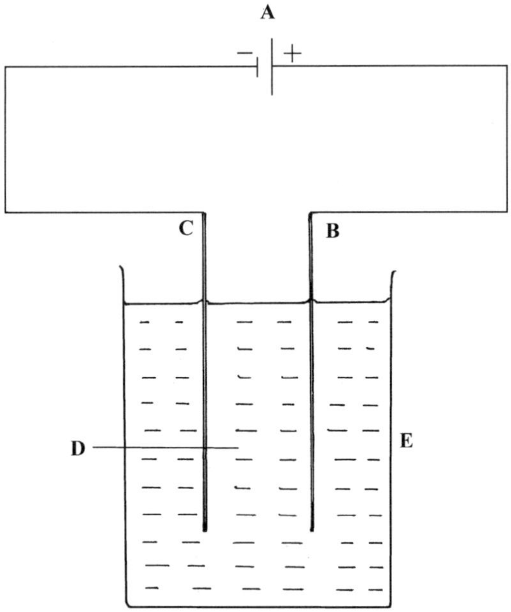

KEY

A = Direct current power supply
B = Anode (+) made of conducting material
C = Cathode (-), as for the anode
D = Electrolyte
E = Suitable container, for example a beaker for aqueous solutions

Figure 6. Basic electrolysis circuit.

What happens during electrolysis?

Electrolytes contain ions (see chapter 2). When the current is passed, the positive ions (metallic and hydrogen ions) move towards the cathode (negative). There they are discharged, by gaining electrons. For example:

$$Cu^{++} + 2e^- \rightarrow Cu$$

Negative ions (non-metals and other ions) move to the anode and are discharged, by losing electrons, for example:

$$Cl^- - 1e^- \rightarrow Cl$$
$$Cl + Cl \rightarrow Cl_2$$

Positive ions are called cations, negative ones, anions.

Quantity, or how much?

The quantity of electricity which has flowed is measured in coulombs:

1 coulomb = 1 ampere x 1 second.

This could, for example, also be
1 coulomb = 0.1 ampere x 10 seconds.

- *Law* – The mass (g) of an ion deposited, for example as an element at an electrode, is directly proportional to the number of coulombs passed. Thus if twice the quantity flows, then twice the mass is deposited.

- *Law* – The same quantity of electricity discharges one mole of all ions carrying the same numerical charge, at an electrode.

Two examples

$$Na^+ + 1e^- \rightarrow Na \text{ (1 mole, 23g)}$$

$$Cl^- - 1e^- \rightarrow Cl \text{ (1 mole, 35.5g)}$$

This quantity is known as the Faraday (F), and is very large, about 96,500 coulombs (after Michael Faraday).

The Faraday corresponds to the flow of a mole of electrons. The discharge of all ions, in molar quantities, takes place in multiples of the Faraday. So, to deposit 1 mole of an ion with a double charge, 2F are required. For example:

$$Mg^{++} + 2e^- \rightarrow Mg \text{ (1 mole, 24g)}$$

Industrial applications of electrolysis

Extraction of sodium metal/chlorine

The electrolyte is fused (molten) sodium chloride, not brine. This is because the sodium reacts with the water as soon as it is discharged at the cathode.

At the cathode (steel)	Na^+	+	$1e^-$	→	Na
At the anode (carbon)	Cl^-	–	$1e^-$	→	Cl
	Cl	+	Cl	→	Cl_2

The sodium metal and chlorine have to be kept apart, otherwise they would react at once reforming sodium chloride and sodium hypochlorite.

Electroplating

In silver plating, the electrolyte is a cyanide of silver. The object to be plated is the cathode, and the anode is a small silver bar.

At the cathode	Ag^+	+	$1e^-$	→	Ag
At the anode	Ag	–	$1e^-$	→	Ag^+

Silver then effectively moves from anode to cathode, the electrolyte remains the same. In the refining of copper the cathodes are thin sheets of pure copper, the anodes are slabs of crude copper. The electrolyte is copper sulphate.

The extraction of aluminium

The electrolyte is purified aluminium oxide, from the ore bauxite. To this is added cryolite, an aluminium fluoride. The idea is to reduce the temperature at which the electrolyte can be kept molten, and so reduce the very large energy consumption of the process.

The cathode is the carbon which lines the large tank in which the electrolysis takes place. The anodes are a series of large carbon rods. These rods burn away quickly and have to be regularly replaced. The aluminium metal forms at the cathode.

The energy consumption of the process is enormous, and forms a major cost of the production. Despite the abundance of other aluminium compounds in the Earths crust, no cheap way of extracting aluminium from these has yet been devised.

At the cathode: $Al^{+++} + 3e^{-} \rightarrow Al$

At the anode: $O^{--} - 2e^{-} \rightarrow O$

$O + O \rightarrow O_2$

The production of sodium hydroxide

This important industrial chemical is produced on a large scale by the electrolysis of brine (sodium chloride solution).

Salt is plentiful and relatively cheap. Various types of cell are used. Hydrogen is produced at the cathode, and chlorine at the anode. The electrolyte gradually changes to aqueous sodium hydroxide. (Compare the manufacture of sodium.)

It is important that the chlorine formed – which is a valuable by-product – is not allowed to mix with the sodium hydroxide. Otherwise a reaction takes place resulting in the formation of sodium hypochlorite, or household bleach. Thus if this is required the chlorine is allowed to mix with the electrolyte as it is formed.

Oxidation and reduction in electrolysis

The modern definition of this is in terms of electron transfer. Remember the word LOGR (Loss of electrons, Oxidation, Gain, Reduction). Thus any chemical species losing electrons is oxidised;

any chemical species gaining electrons is reduced.

For example, sodium burns in chlorine to form sodium chloride:

$$Na - 1e^{-} \rightarrow Na^{+}$$

$$Cl - 1e^{-} \rightarrow Cl^{-}$$

The sodium is oxidised (loss), and chlorine reduced (gain). Any species losing electrons is therefore a reducing agent, whilst those gaining electrons are oxidising agents.

► *Study tip* – Remember that those elements on the left of the Periodic Table (metals and hydrogen), are usually reducing agents. Those on the right (non-metallic elements) are oxidising agents. However, there are plenty of other ions and substances which are oxidising and reducing agents apart from elements (see also chapter 3).

Example of an oxidising ion

A simple example of an oxidising ion is the permanganate ion. This contains the element manganese with a plus 7 oxidation number (see chapter 3). When reduced it finishes up with a lower oxidation number of plus 2, the normal oxidation state of manganese (Mn^{++}).

Example of a reducing ion

A simple example of a reducing ion is the iron (II) ion. This is fairly easily oxidised, by loss of one electron, to Fe (III). It is important to understand that the two processes, oxidation and reduction, always occur together.

$$Fe^{++} - 1e^{-} \rightarrow Fe^{+++}$$

Key points to understand and learn

1. Electrolysis, electrolyte, ion (see chapter 2) anode and cathode, and what happens at the electrodes.
2. The laws of electrolysis and industrial applications.
3. Oxidation and reduction. Simple calculations in electrolysis.

A worked example

The following apparatus is used in the electrolysis of fused lead chloride.

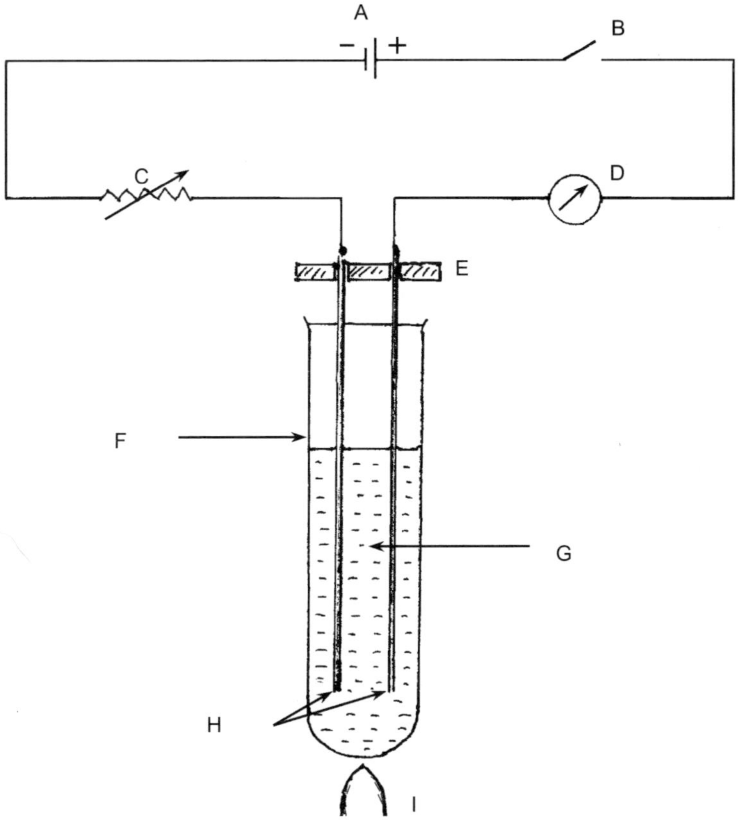

KEY

A Battery power supply
B Switch
C Rheostat for steady current
D Ammeter
E Electrode holder
F Large hard-glass test-tube
G Molten electrolyte
H Carbon rod electrodes
I Bunsen burner flame to keep electrolyte molten

NOTE

Lead bromide can be used instead of lead chloride

Figure 7. The electrolysis of fused lead chloride

A stop watch is also required.

In the experiment, a current of 2A was passed for 4 hours, when 30.9 g of lead was deposited at the cathode. What is the charge on the lead ion? (1 Faraday = 96,500 coulomb, and the relative atomic mass of lead = 207).

Step 1
Total number of coulombs passed:

= current x time (seconds)
= 2 x 60 x 60 x 4
= 28,800 coulomb

Step 2
y x Faraday's deposit 207g of lead.

In other words,

y x 96,500 = 207g, where y is the charge on the lead ion.

Step 3
In the experiment, 28,800 coulomb deposits 30.9 g of lead. Thus,

$$y \times \frac{96{,}500}{28{,}800} = \frac{207}{30.9}$$

In other words,

$$y = \frac{(207 \times 28{,}800)}{(96{,}500 \times 30.9)}$$

from which y = 2

Thus the charge on the lead ion is 2 plus. It follows that the formula for lead chloride is $PbCl_2$, since the chloride ion is Cl^-.

The 'water' voltameter

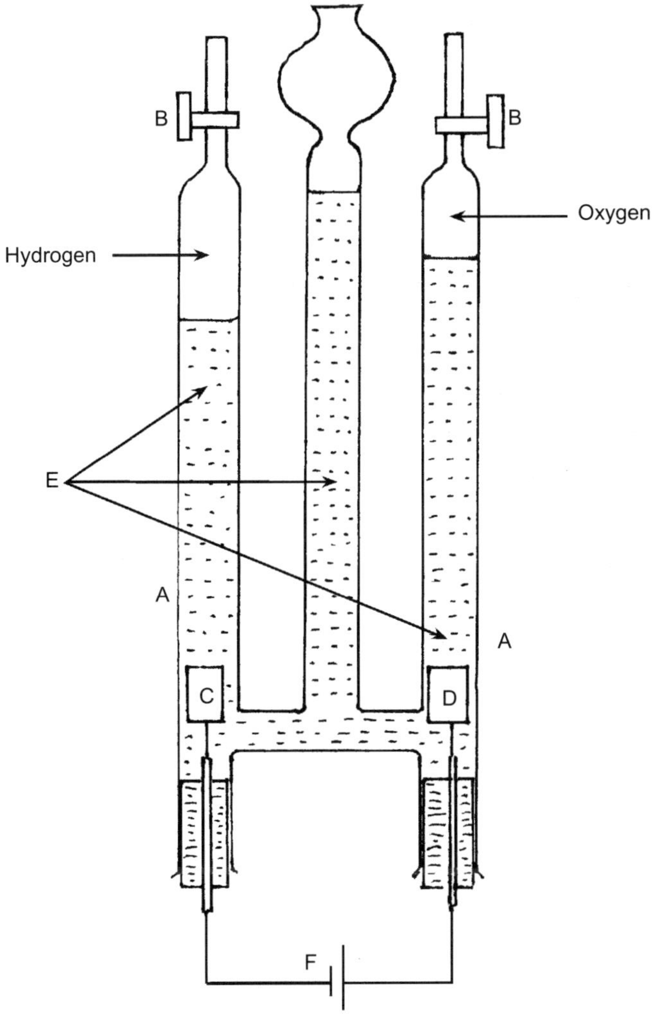

KEY

A The water voltameter
B Taps to release gases
C Platinum cathode
D Platinum anode
E Dilute sulphuric acid
F Power supply

Figure 8. The water voltameter

This special piece of apparatus is used to show the electrolysis of water. The electrolyte is dilute sulphuric acid, which is a good conductor. The electrodes are platinum. The net effect of the experiment is the electrolysis of water.

The two inverted graduated tubes measure the volume of hydrogen produced at the cathode and oxygen at the anode. It is found that two volumes of hydrogen are produced for every one of oxygen. Since these volumes are directly convertible to the atomic ratio of hydrogen to oxygen in water, it follows that the formula for water is H_2O.

Case study: the persistent chemist

Many chemists had wrestled with the isolation of the element fluorine. It heads Group VII in the Periodic Table, and is the most reactive of all the non-metallic elements and possibly the most reactive of all the elements. It had successfully withstood all efforts to isolate it as the element.

Then, in the 1880s, the French chemist Henri Moissan decided that an electrolytic method was the only likely approach, after others had failed. First of all, he had tried the electrolysis of aqueous solutions of hydrofluoric acid HF. This however only produced hydrogen at the cathode and oxygen at the anode – usually the case with dilute acids. Moissan then tried pure hydrofluoric acid, and found this to be non-conducting. However, by dissolving potassium fluoride in the pure acid he obtained a conducting solution and for the first time the element fluorine was seen. It is a gas resembling chlorine.

Helping you learn

Progress questions

1. Calculate the amount of copper deposited at the cathode by a flow of 1 amp for 4 hours. (RAM of copper = 63.5, and 1 Faraday of electricity = 96,500 coulomb.)

2. Write ionic half equations for MnO_4^- being reduced to Mn^{++}

by gaining electrons and the oxidation of iron (II) to iron (III) by losing electrons. Now combine the two half equations correctly to eliminate the electrons. In the first equation you will need to include hydrogen ions, that is H^+. Remember, in ionic equations both the atoms and charges on both sides of the equation must be *balanced.*

Discussion points

1. Pure (concentrated) sulphuric acid is almost a non-conductor, whilst dilute sulphuric acid is a very good one. When concentrated sulphuric acid is diluted, a large amount of heat is generated. What can you say about the structures of both substances. Should the same formula, H_2SO_4, be applied to both?

2. Electrolytes contain ions. Generally substances which do not conduct do not contain ions, even though they may be soluble in water. A good example is sugar, which dissolves easily in water but is a non-electrolyte. What type of bonds, therefore, do you think sugar contains?

Practical assignments

1. Investigate the electrolysis of dilute hydrochloric acid, and then concentrated hydrochloric acid. Use a suitable beaker as an electrolytic cell, a 6v power supply and carbon rod electrodes. Take great care when using concentrated hydrochloric acid, because it is corrosive.

2. Investigate the conductivity of sugar solution using the circuit shown in diagram 1. Include a bulb to show any conduction through the solution. Next, using the same circuit, with graphite electrodes, investigate how the conductivity of dilute sulphuric acid changes with increasing dilution. Start with ordinary bench dilute sulphuric acid, which is about 2 M strength, and dilute in equal steps. If you substitute an ammeter for the bulb, and take readings of the current in the circuit each time, you should then plot a graph of the results.

Study and revision tips

1. Make a list of all the important terms used in electrolysis. Make sure you understand and learn them.

2. Understand and learn the two laws of electrolysis.

3. Make sure you know how to write half and full ionic equations, and how to balance them. You will need to learn the formulae for the more common ions.

4. Revise the most important industrial applications of electrolysis. These include the large-scale preparation of chlorine, the extraction of aluminium and the refining of copper.

7

Classifying Compounds and Reactions

One-minute summary – In this important chapter you will learn about how chemical compounds are classified, and how most reactions in chemistry fall into a pattern. This helps us to see the whole subject as an orderly one, and not just as masses of unrelated facts to be learned. After the classification of compounds and reaction type, we look at ionic equations. Next we look at tests for the most frequently found cations and anions. Finally there are some exercises for you to practise. In this chapter we will explore:

- the classification of chemical compounds
- the classification of reactions in chemistry
- tests for ions in solution.

The classification of chemical compounds

The broadest subdivision of chemical compounds is into inorganic and organic compounds.

Organic compounds

Organic compounds are those containing the element **carbon**. A few compounds – for example carbon dioxide and carbonates – are regarded as inorganic. Here are some examples of organic compounds:

ethanol
benzene
sugar
vitamins

Organic chemistry is dealt with in detail in a later chapter.

Inorganic compounds

Those compounds formed between the metals on the left-hand side of the Periodic Table and the non-metals on the right side are inorganic and also a fairly large number of compounds formed between non-metallic elements.

Inorganic compounds consist mainly of acids, bases, and salts.

1. All acids contain hydrogen, as ions in dilute aqueous solution.

2. Bases have the ability to neutralise acids. An alkali is simply a soluble base – sodium hydroxide is an example. The oxides of most metals are bases.

3. A salt is formed when acids and bases react together to form a neutral solution. Most metals form salts. They are generally soluble in water, but there are some important exceptions that are used as tests for ions in solution.

Table 8 shows the main properties of acids, in tests. These tests apply to dilute acids, not concentrated acids.

Test	*Reaction given by acids*
With blue litmus	Goes red.
With metals	Gives hydrogen, if the metal is above hydrogen in the reactivity series.
With alkalis/bases	Neutralises the acid, water formed.
With carbonates	Fizzes, carbon dioxide produced.
Electrical conductivity	Generally good conductors, they are strong electrolytes.

Table 8. The behaviour of acids.

All the properties shown in table 8 are due to the presence of hydrogen ions in all acids. In addition, each acid gives the reactions of the particular anion. For example, dilute hydrochloric acid gives the reactions of the chloride ion. This is discussed later.

Alkalis and bases neutralise acids

The main properties of alkalis and bases are that they neutralise acids to form salts, and turn red litmus blue. Common examples are sodium and potassium hydroxides (alkalis), and calcium and magnesium oxides (bases).

Non-metals in the Periodic Table combine with each other as mentioned above, but the compounds formed do not contain ions. They are therefore covalent in their bonding. Phosphorus, for example, forms two chlorides of formulae PCl_3 and PCl_5; both show the properties of covalent compounds.

The classification of reactions in chemistry

Table 9 shows all the main types of reactions. We will look at each of these in turn.

	Type of reaction
1	thermal decomposition
2	synthesis
3	precipitation
4	combustion
5	neutralisation
6	oxidation and reduction (chapter 6)
7	reversible reactions

Table 9. Types of chemical reaction.

Thermal decomposition

Thermal decomposition takes place when a single substance, as a solid, is heated in a special heat resistant test-tube and breaks down to give two or more simpler products. In other words, the reaction is of the form:

$$A \underset{\text{heat}}{\rightarrow} B + C$$

Some examples are:

Nitrates

These generally decompose as follows. The example shows lead nitrate:

$$2Pb(NO_3)_2\ (s) \rightarrow 2PbO\ (s) + 4NO_2\ (g) + O_2\ (g)$$

The second product, nitrogen dioxide, is a brown toxic gas. The nitrates of sodium, potassium and ammonium give different products.

Carbonates

These produce carbon dioxide and leave a metallic oxide. For example:

$$MgCO_3\ (s) \rightarrow MgO(s) + CO_2(g)$$

The carbonates of sodium and potassium in Group I are not normally decomposed; those of the Group II metals below magnesium require high temperatures.

Sulphates

Most sulphates are decomposed, on strong heating, to leave the oxide of the metal and give off sulphur dioxide (SO_2) and/or sulphur trioxide (SO_3) as gases.

Halides

These are not normally affected, although those of ammonium sublime ie turn into vapour at once.

Organic solid compounds

These compounds are not stable to heat. Even at moderate temperatures, they break down to leave carbon and a complex variety of other products.

Synthesis

Synthesis is the making of a compound, or compounds, from simpler starting materials. The process is very important in the chemical industry, for example in the large-scale manufacture of ammonia and sulphuric acid.

Ammonia NH_3
The raw materials are pure nitrogen gas (from liquid air) and pure hydrogen. The reaction equation is:

$$N_2\ (g)\ +\ 3H_2\ (g)\ \underset{400^\circ C/\text{catalyst}}{\text{200Atm pressure}} \rightleftharpoons 2NH_3\ (g)$$

The sign $\rightleftharpoons$ indicates that the reaction is reversible. The best conditions to promote the formation of ammonia are high pressure (200 atm) and a temperature of 400 – 450°C. The starting gases must be pure, since impurities dramatically reduce the effectiveness of the catalyst. This is referred to as poisoning of the catalyst.

Sulphuric acid H_2SO_4
The raw materials are pure sulphur dioxide (SO_2) and pure oxygen. The sulphur dioxide is obtained by burning sulphur in air:

$$S + O_2\ (g) \rightarrow SO_2\ (g)$$

The oxygen is obtained from liquid air. The two are mixed, and passed over a catalyst at 500°C, when sulphur trioxide is formed. The equation is:

$$2SO_2\ (g) + O_2\ (g) \rightarrow 2SO_3\ (g)$$

The starting gases in the above equation must be pure for the same reason as in the synthesis of ammonia; an impure product would result. The sulphur trioxide is not dissolved in water, but in concentrated sulphuric acid, which absorbs it more easily. Water is then added and so on. The overall reaction is:

$$SO_3\ (g) + H_2O\ (l) \rightarrow H_2SO_4\ (l)$$

Precipitation and ionic equations
This reaction is commonly used to test for:

1. halide ions (Cl^-, Br^- and I^-) in aqueous solution

2. the sulphate ion SO_4^{--}, the carbonate ion CO_3^{--}

3. the ions of some metals.

► *Precipitation* – occurs when two substances, usually salts, in aqueous solution are mixed, and a solid insoluble substance, or precipitate, is immediately formed.

For example when solutions of silver nitrate and a soluble chloride are mixed, a white precipitate of silver chloride is immediately formed. Here is the equation:

$$AgNO_3\ (aq) + NaCl\ (aq) \rightarrow AgCl\ (s) + NaNO_3\ (aq)$$

If we write out both starting substances as their ions, the following equation is obtained:

$$Ag^+\ (aq) + NO_3^-\ (aq) + Na^+\ (aq) + NO_3^-\ (aq) \rightarrow$$

$$AgCl\ (s) + Na^+\ (aq) + NO_3^-\ (aq)$$

If we remove the ions common to both sides of this equation (Na^+ and NO_3^-) – since they are there at the start and finish of the reaction – the equation reduces to:

$$Ag^+\ (aq) + Cl^-\ (aq) \rightarrow AgCl\ (s)$$

The ions that have been eliminated are sometimes referred to as 'passenger ions'. *Any soluble chloride gives the same result*, including dilute hydrochloric acid. Hence its use to demonstrate the test for the chloride ion. Bromides (Br^-) and iodides (I^-) give similar results.

The above equation is an example of an **ionic equation**, and there are many others.

The test for the sulphate ion

A similar precipitation reaction takes place in the test for the sulphate ion. The sulphate ion, in aqueous solution from any salt, when added to barium chloride solution, gives an immediate precipitate of white barium sulphate. For example, with the salt sodium sulphate:

$$BaCl_2\ (aq) + Na_2SO_4(aq) \rightarrow BaSO_4\ (s) + 2NaCl\ (aq)$$

Written out ionically:

$$Ba^{++}(aq) + 2Cl^-\ (aq) + 2Na^+\ (aq) + SO_4^- \rightarrow$$

$$BaSO_4\ (s) + 2\ Na^+\ (aq) + 2\ Cl^-\ (aq)$$

which reduces to:

$$Ba^{++}\ (aq) + SO_4^{--}(aq) \rightarrow BaSO_4\ (s)$$

All soluble sulphates, including dilute sulphuric acid, give this reaction. Here are two exercises for you to try. As before, ask for the help of your teacher to obtain the chemicals needed and a supply of test tubes.

Exercise 1

Obtain a solution of strontium nitrate (or chloride) in water. Strontium lies between calcium and barium in Group II of the Periodic Table. Divide your original solution into four portions. Add dilute sulphuric acid to one of your four portions. Repeat the experiment with solutions diluted by approximately a factor of two, three and four times. Notice the formation of the precipitate of strontium sulphate. Write an ionic equation. Can you explain the results?

Exercise 2

Perform the chloride ion test, using potassium chloride. Then repeat the same test for the bromide and iodide ions, using solutions of potassium bromide and iodide respectively. Note the colours of the three precipitates. Can you explain the change in the colour of the precipitate of silver chloride after a minute or two?

Combustion

Combustion is the same as burning. It is perhaps the most familiar reaction to students starting out on the subject. Combustion takes place when any flammable material burns in oxygen (or air). It can apply to other reactions, too, not involving oxygen. The reaction involves a flame and light.

All fuels, other than nuclear fuels, rely on combustion in air to produce their energy, for example coal, oil, methane, and petrol. The human race has learned how to harness this energy very successfully to generate electricity, and to power all kinds of transport.

Acetylene (ethyne -C_2H_2) produces a very high temperature on combustion with oxygen, and is widely used in cutting and welding metals.

Neutralisation and the pH scale

Neutralisation takes place between any alkali (a soluble base) or base, and any acid. Here is a familiar example:

$$2NaOH\ (aq) + H_2SO_4\ (aq) \rightarrow Na_2SO_4\ (aq) + 2H_2O\ (l)$$

This equation is written out ionically, since it takes place in an aqueous solution:

$$2Na^+(aq) + 2OH^-(aq) + 2H^+(aq) + SO_4^{--}aq) \rightarrow 2Na^+(aq) +$$

$$SO_4^{--}(aq) + 2H_2O\ (l)$$

We can cancel out the Na^+ and SO_4^{--} ions. (These types of ion, as mentioned before, are common to both sides in an ionic equation, and are sometimes referred to as passenger or spectator ions.) The equation reduces to:

$$H^+\ (aq) + OH^-\ (aq) \rightarrow H_2O\ (l)$$

This simple ionic equation is the basic reaction taking place *in all neutralisations* between any acid and any alkali.

Chemists distinguish between strong and weak acids. These are not the same as concentrated and dilute acids.

- *A strong acid* – provides, in an aqueous solution, a very high concentration of H^+ ions, this ion being solely responsible for making a solution acidic. This is why strong acids are very good conductors.

- *Weak acids* such as ethanoic acid (an organic acid) provide far less H^+ ions and are less good conductors of a current.

The pH scale

This scale was devised to give a numerical value to the amount of acidity or alkalinity of a solution. The scale extends from 0 to 14, 7 being exactly neutral.

1. Values of pH below 7 are acidic. The lower the value, the more acidic.

2. Values above 7 are alkaline. The higher the value, the more alkaline is the solution. Figure 9 shows a diagram of the pH scale.

Colours shown with litmus solution

< Increasingly acidic Increasingly alkaline >

Neutral

pH <————————————————> pH

0 1 2 3 4 5 6 7 8 9 10 11 12 13 14

Litmus goes red Litmus goes blue

Colours shown with universal indicator

<increasingly acidic Increasingly alkaline >

| 7 |

< 0 5 Neutral 9 14 >

Colours (below) shown with universal indicator

<——————> <——————> <——————> <——————>

0 Red 5 Orange/ Yellow 7 GREEN Blue/ Green 9 Blue – Purple 14

Figure 9. The pH scale.

A rough idea of the pH value of any solution may be estimated by adding a little universal indicator. This shows a range of different colours over the whole pH scale range, giving an idea of the pH of an unknown solution. More accurate methods involve using a pH meter. The pH scale is important to farmers and growers who sometimes need to know the pH of a soil.

Oxidation and reduction

We have already dealt in detail with this topic, but here is a brief summary. Note the following:

1. The two processes *always* occur together – one species acting as oxidising agent, the other acting as reducing agent.

2. When a species *loses* electrons it has been oxidised. It is therefore a reducing agent. When a species *gains* electrons it has been reduced and so acts as an oxidising agent. (Remember the word **LOGR**.)

3. Metals and hydrogen are reducing agents especially the Group I and Group II metals, whilst non-metals – especially those of Groups VI and VII – are oxidising agents. Caesium (Group I) is theoretically the strongest reducing agent, and fluorine (Group VII) the strongest oxidising agent.

4. There are other ions and radicals which are either oxidising or reducing agents. For example the Fe^{++} (aq) ion is a reducing ion. The MnO_4^- (aq) ion is a strong oxidising ion.

Reversible reactions

A reversible reaction is any chemical change which, by altering the physical conditions, or the concentration of one or both components of the reaction, can be made to proceed in *either* direction. $\rightleftharpoons$ denotes a reversible reaction.

The best examples are found in gaseous reactions, but there are plenty of others. A common example is the driving off – by heating – of water of crystallisation. Thus copper sulphate is normally met with as blue crystals, with formula $CuSO_4.5H_2O$. The 5 moles of water are the 'water of crystallisation'.

$$CuSO_4.5H_2O\ (s)\ \text{strong heat} \rightarrow CuSO_4\ (s) + 5H_2O(l)$$

The starting substance takes the form of blue crystals whilst the anhydrous (without water) form on the right hand side of the equation is white. When water is added to this, heat is given out, and it reverts to the blue hydrated form, in other words:

$$\underset{\text{white}}{CuSO_4\ (s)} + 5H_2O\ (l)\ \text{room temp.} \rightarrow \underset{\text{blue}}{CuSO_4.5H_2O\ (s)}$$

The overall equation is:

$$CuSO_4.5H_2O\ (s) \rightleftharpoons CuSO_4\ (s) + 5H_2O(s)$$

Tests for ions in solution

Aqueous metallic ions

Most 'heavy metal' (other than Groups I and II) ions in aqueous solution can be detected by making a solution in water, and then adding sodium hydroxide solution. For most metals, a precipitate of an insoluble hydroxide is formed. For example, for iron (II), Fe^{++} (aq) ions:

$$Fe^{++}\ (aq) + 2OH^-\ (aq) \rightarrow Fe(OH)_2\ (s)$$

Most of these precipitates are coloured. Table 10 shows the colours.

Metal ion	*Colour of precipitate*	*Soluble in excess NaOH*
Mg^{++}	white	No
Al^{+++}	white	Yes
Fe^{++}	darkish green	No
Fe^{+++}	red-brown	No
Pb^{++}	white	Yes
Zn^{++}	white	Yes
Cu^{++}	blue	No

Table 10. Colours of precipitates.

The precipitates formed by Al^{+++}, Pb^{++} and Zn^{++} are all white, but may be distinguished by additional tests.

The ions of Groups I and II

The ions of Groups I and II are normally distinguished by **flame tests.** A *clean* loop of wire is dipped into hydrochloric acid, and then into the solid substance. The loop is then held in a **blue** (hot) Bunsen burner flame. Group I and II cations give the colours shown in table 11. (Not Mg^{++}, see the previous table).

Cation	*Colour*
Li^{+}	crimson
Na^{+}	yellow
K^{+}	lilac
Ca^{++}	brick red
Sr^{++}	deep red
Ba^{++}	apple green

Table 11. Colours produced by the Bunsen burner test.

You may have noticed that one cation is not covered by either of these two tables. This is the ammonium ion NH_4^{+}. It is easily tested for as follows. Add bench sodium hydroxide to some of the solid in a test-tube solution and warm. The distinctive smell of ammonia gas very soon becomes apparent. Test for this with moist red litmus paper, which goes blue, since ammonia is an alkaline gas.

Anions

A solution of the solid substance is made in distilled water, except when testing for carbonates and bicarbonates. The common anions tests are shown in table 12.

Anion	*Test*	*Result*
Chloride Cl^-	Add a little dilute nitric acid and silver nitrate solution.	White precipitate discolours.
Bromide Br^-	As above.	Very pale yellow precipitate.
Iodide I^-	As above.	Pale yellow precipitate.
Carbonate CO_3^{--}	Add dilute hydrochloric acid to the solid.	Fizzes and CO_2 formed.
Bicarbonate HCO_3^-	As for carbonate.	As for carbonate.
Sulphate SO_4^{--}	Add dilute hydrochloric acid, then barium chloride.	Instant white precipitate.
Nitrate NO_3^-	Add a little clean aluminium foil, bench sodium hydroxide and warm gently.	Ammonia is formed. Test with red litmus.

Table 12. Tests for anions. ('Radicals')

Tests for gases can be found at the end of Chapter 11.

Helping you learn

Progress questions

1. Hydrated magnesium nitrate has the formula $Mg(NO_3)_2.6H_2O$. What will happen when it is heated until no further change occurs? Write a balanced equation to show the changes.

2. A white solid was soluble in water. Its solution gave white precipitates with acidified silver nitrate solution and also with dilute sulphuric acid. What might the solid be? Outline further tests to check your ideas.

3. Write an ionic equation for the precipitation of lead iodide, formula PbI_2, from aqueous lead nitrate solution and potassium iodide solution. The precipitate of lead iodide is yellow. This precipitate formed dissolves if the mixture is boiled, producing a colourless solution. Can you explain this observation?

4. Classify, from the list at the beginning, each of the following reactions represented by an equation for each (there may be more than one answer).

   a. $BiCl_3\ (l) + H_2O\ (l) \rightleftharpoons BiOCl(s) + 2HCl(aq)$

   b. $Sn^{++}\ (aq) + 2Fe^{+++}\ (aq) \rightarrow Sn^{++++}\ (aq) + 2Fe^{++}\ (aq)$

   c. $2Cu^{++}\ (aq) + 4I^{-}\ (aq) \rightarrow 2Cu_2I_2\ (s) + I_2\ (s)$

   d. $2KNO_3(s) \rightarrow 2KNO_2\ (s) + O_2(g)$

Discussion points

Imagine you are an industrial chemist, wishing to produce the substance C on a large scale, from the starting substances A and B. C is a gas at temperatures above 100°C. The reaction can be represented by the equation:

$$2A(g) + B(g) \rightarrow 2C(g)$$

A and B, like C are gases at temperatures above 100°C. However, A and B are 'permanent' gases i.e. have low boiling points, but the product C becomes a solid at 30°C. Answer the following questions about the process from the above information.

1. What volume change occurs at 100°C, if all of A and B, as in the equation above, are converted to C?

2. In the light of your answer to question 1 above, do you think the left to right change is favoured by an increase or a decrease from normal atmospheric pressure in the process?

3. The left to right reaction is known to be exothermic. What temperature conditions should therefore favour the formation of C?

4. In view of the fact that C becomes a solid at 30°C what precaution might you have to take in the plant being used to manufacture C?

5. A 'finely divided' metallic catalyst is found to promote the formation of C in the reaction. What is meant by 'finely divided', and why is the catalyst in this state?

6. The catalyst used is known to be prone to reduced efficiency by 'poisoning'. Explain ways in which this problem may be got round.

Practical assignments

1. Perform the two experiments outlined in section 3, if you have not already done so.

2. Attempt the two experiments with hydrated copper sulphate $CuSO_4.5H_2O$ and magnesium nitrate $Mg(NO_3).6H_2O$ to confirm your suggestions in the exercises above. They should be conducted under the supervision of your teacher.

3. Ask your teacher to give you an unknown solid substance for identification. Carry out the tests on the substance using the tables of tests for cations and anions set out in this chapter.

Study and revision tips

1. This topic is still popular with examiners. Questions of the type given in the exercises occur frequently. It is therefore particularly important that you attempt them. Check with the answers at the end of the chapter. The tests for cations and anions are frequently required for examination questions.

2. Make sure that you understand and know the different types of reaction in chemistry and at least one example of each. Ionic equations are important and popular with examiners, especially the special case of neutralisation. It is a good idea to practice ionic equations by choosing any precipitation reaction and eliminating the common ions. You should be fully conversant with the tests for common cations and anions.

8

Rates of Chemical Change

One-minute summary – The 'rate of a chemical change' means the speed at which a particular reaction takes place, under a given set of conditions. It can be very important to industrial chemists: they generally wish to manufacture a product in as short a time as possible. Rates of reaction vary from virtually instantaneous to slow. Examples are the precipitation of barium sulphate ($BaSO_4$) in the test for the sulphate ion:

$$H_2SO_4\ (aq) + BaCl_2\ (aq) \rightarrow BaSO_4\ (s) + 2HCl\ (aq)$$

This reaction is almost instantaneous. On the other hand, the hardening of mortar and cement, and the rusting of iron are slow. Here is a simple explanation. Lets assume that, for reaction to take place, two molecules or ions have to collide. The factors which speed up rates of chemical change are those which promote the greatest number of collisions. Thus, by bringing the particles closer together – for example by increasing the pressure in a gaseous reaction, or giving them extra energy by raising the temperature – more collisions take place. In this way the rate of the change is increased. In this chapter we will learn about:

- factors which affect reaction rate
- experimental investigation
- using mass loss to follow a reaction rate
- the reaction between metal and dilute acid.

Factors which affect reaction rate

1. *Temperature* – an increase normally speeds up a reaction.

2. *Concentration* – an increase also usually speeds up a reaction.

3. *Particle size* – the smaller the particle size, where one reactant is a solid and the other a liquid, also increases reaction rate.

4. *Change in gas pressure in gaseous reactions* – an increase in pressure usually speeds up a gaseous reaction.

5. *The use of a catalyst* – another chemical is employed to speed up a reaction, but takes no part in the chemical change. Catalysts are important in some manufacturing processes that involve gases.

Experimental investigation

Various experimental methods are available. The choice depends on what sort of reaction we wish to investigate.

A popular choice to illustrate 'reaction rate' is the reaction between sodium thiosulphate solution and dilute acid. This is because the effects of both concentration and temperature can be readily investigated. In the reaction, a precipitate of sulphur is gradually produced, making the reaction mixture cloudy.

The reaction is normally carried out in a conical flask, standing on a piece of paper with a large dot or cross marked on it. The reaction time is that between when the thiosulphate solution and dilute acid are mixed, to when the dot or cross disappears from view. Here is the reaction:

$$S_2O_3^{--}\,(aq) + 2H^+\,(aq) \rightarrow S\,(s) + SO_2(g) + H_2O\,(l)$$

As stated above, this reaction can easily be used to investigate both temperature and concentration changes. We can plot graphs of both sets of results.

► *Note* – It is very important to keep all other variables constant.

The same person should decide when the mark has disappeared, and the same depth of liquid must be looked through all the time. The interval between the initial mixing and the point where the mark disappears must be measured with a stop watch. Similar measurements should be taken with different concentrations of acid.

In the second experiment, take readings at increasing temperatures, from (say) 20°C to 70°C giving six readings. You can then plot graphs of both sets of results.

▶ *Conclusion* – The experiment shows that the rate of reaction increases at higher temperatures. It also increases if the concentration of either of the reactants is increased.

Using mass loss to follow a reaction rate

When a dilute acid (hydrochloric or nitric acid, not sulphuric) is added to powdered calcium carbonate, carbon dioxide (gas) is evolved. Since the carbon dioxide escapes from the reaction environment, the whole loses mass. The rate at which this escape of gas takes place tells us how quickly the reaction proceeds.

The experiment

Place the flask and calcium carbonate on a (direct reading) balance, weigh them initially and then add the dilute acid. Then take mass readings at regular intervals, for example every 20 seconds. You can then plot a graph of weight loss against time. You can use the same method to investigate particle size, for example using calcium carbonate in the form of small lumps as well as temperature changes and concentration of the acid.

▶ *Conclusion* – The results will show that the reaction is quickest with powdered calcium carbonate. As with the experiments with sodium thiosulphate, the rate increases at higher temperatures and with increased concentration of acid.

Figure 10 shows a graph of the results. The graph is steepest at the beginning and then flattens off, as the reactants are used up.

$$CaCO_3(s) + 2HCl(aq) \rightarrow CaCl_2(aq) + H_2O(l) + CO_2(g)$$

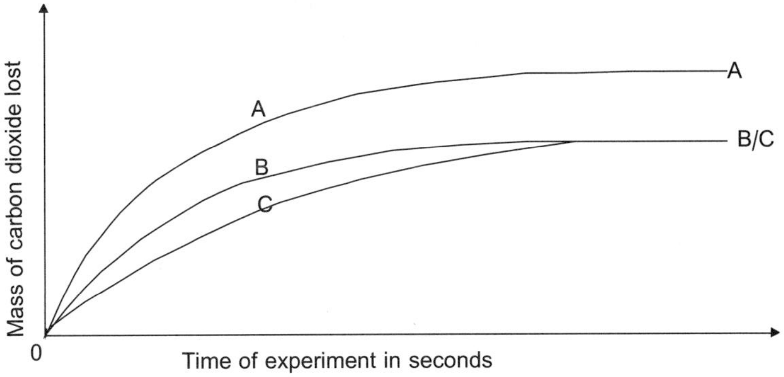

KEY
A (top graph) - powdered calcium carbonate, showing the greatest rate loss.
B (middle graph) – small lumps of calcium carbonate
C (lower graph) – larger lumps of calcium carbonate

NOTES
1. In all three experiements an excess of dilute acid is used.
2. In A, with the higher flat portion, a greater mass of calcium carbonate was used initially (same masses in B and C).

Figure 10. Graphs showing mass loss against time in seconds in the reaction between dilute hydrochloride acid and calcium carbonate.

The reaction between metal and dilute acid

Using a gas syringe to measure the volume of hydrogen produced
The method described here is suitable where small volumes are involved. See figure 11 for a diagram of the apparatus.

Items of apparatus required
1. tap funnel to deliver the acid
2. conical flask and contents (e.g. magnesium/dilute acid)
3. gas syringe to collect and measure the volume of gas.

Read off at equal intervals the volume of hydrogen produced, for example every 20 seconds.

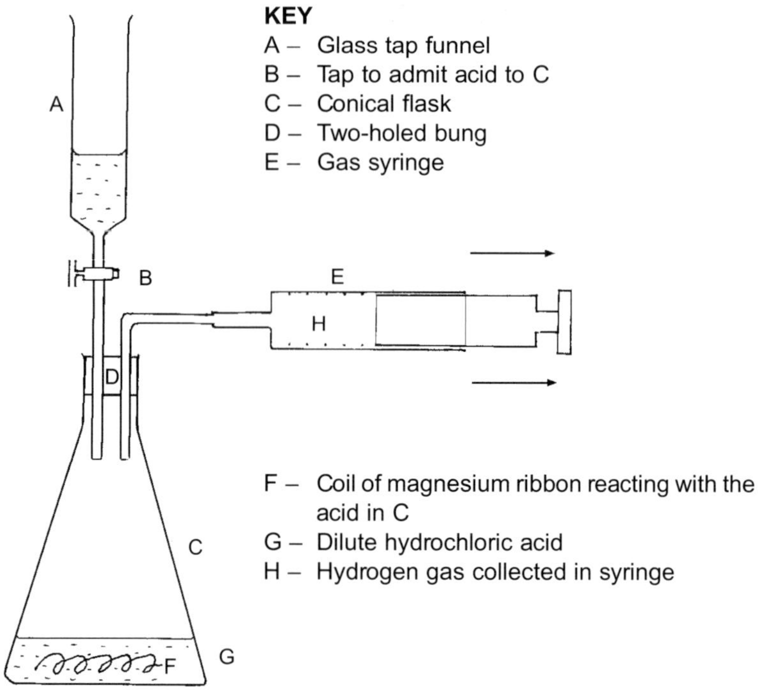

Figure 11. Apparatus to follow the reaction between metal and dilute acid using a syringe to collect and measure the volume of hydrogen produced.

Helping you learn

Progress questions

1. Describe, using the terms fast, moderate or slow, the rate of each of the following:

   (a) potassium reacting with cold water

   (b) calcium reacting with cold water

   (c) neutralising acid with alkali

   (d) the rusting of a piece of galvanised iron

(e) the precipitation of silver chloride in the test for a chloride in aqueous solution.

3. The graph in figure 12 (the top graph) shows the total volume of hydrogen produced by reaction between a piece of magnesium ribbon and excess dilute sulphuric acid. Answer the following questions:

(a) How long did it take to produce 30 cm^3 of hydrogen?

(b) What volume of hydrogen was produced after 40 seconds?

(c) What volume of gas was produced after 80 seconds?

(d) What does the flat portion of the graph represent?

(e) Mark on the graph the volume of hydrogen you would get if the same mass of magnesium had been used, but starting with more concentrated acid.

(f) What experimental changes would the lower graph show?

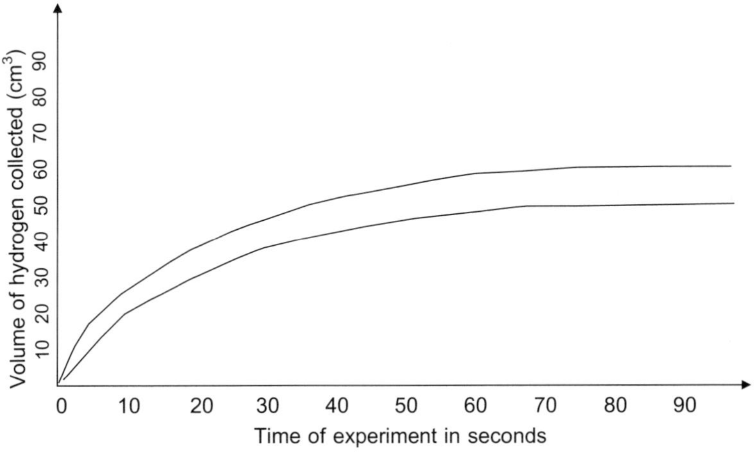

Figure 12. Graphs of reaction between magnesium and dilute acid.

Discussion points

1. When hydrogen gas and chlorine gas are mixed in dull light, no visible reaction takes place. As soon as the mixture is exposed to direct sunlight, it explodes. Can you explain this?

2. Which five factors can affect reaction rate?

Practical assignments

Devise suitable experiments to investigate:

1. The catalytic effect of powdered manganese dioxide on the decomposition rate of aqueous hydrogen peroxide, according to the equation:

$$2H_2O_2\ (aq) \rightarrow 2H_2O\ (l) + O_2\ (g)$$

2. The catalytic effect of powdered manganese dioxide on the decomposition rate, on heating, of solid potassium chlorate, according to the equation:

$$2KClO_3\ (s) \rightarrow 2KCl\ (s) + 3O_2\ (g)$$

Your account should describe suitable apparatus and the measurements required.

Key points

1. Increased temperature and concentration almost invariably increase reaction rate. A graph of results usually shows a gradually decreasing rate as the concentration of one or both reactants falls as they are used up.

2. Catalysts affect the rate of many reactions. They generally accelerate them.

Study and revision tips

The 'rates of chemical change' is another popular topic with examiners. Make sure you:

1. understand why knowledge of reaction rates is important to the chemical industry

2. can give a simple explanation, in terms of number of collisions, of the speed of a reaction

3. know the major factors that affect the rate of a reaction

4. have a reasonably detailed knowledge of at least one experiment to demonstrate each of the major factors

5. can draw and interpret graphs.

9

Organic Chemistry

One-minute summary – Organic chemistry is the chemistry of **carbon** and its **compounds**. All organic compounds contain carbon, and there are many thousands of them. However, a few simple compounds are still regarded as 'inorganic', such as carbon monoxide (CO), carbon dioxide, and carbonates. Huge numbers of carbon compounds exists, because carbon bonds so easily with hydrogen and oxygen – and with itself. This creates chains of carbon, many of them very long. The **structural formulae** of these these bonds are simply represented by straight lines joining the various atoms in the molecule. Carbon occurs in all organic compounds – oxygen and hydrogen most of the time – and occasionally other non-metals such as nitrogen and sulphur. **Isomerism** further increases the huge range of organic compounds that exist. In this chapter we will explore:

- general characteristics of organic compounds
- functional groups and homologous series
- unsaturated compounds
- fermentation
- polymerisation
- isomerism.

Carbon forms bonds easily with hydrogen and oxygen – and in particular with itself. This gives rise to chains of carbon atoms which can be very long. As well as these long chains, carbon can form compounds with branched chains, and also with rings of carbon atoms, six-membered rings being the commonest. Here we will just look at a few of them.

Carbon has a valency, or combining power, of four. It normally forms four single covalent bonds with other atoms, or groups, in organic compounds. These bonds are simply shown by straight

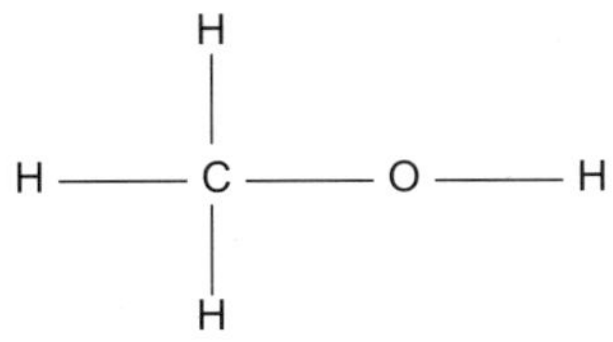

The structural formula for methanol CH_3OH

NOTES

1. H = Hydrogen C = Carbon O = Oxygen
2. ——— represents a bond between two adjacent atoms.
3. The three elements present all show their normal oxidation numbers H – 1, C – 4, O – 2.

Figure 13. The structural formula of methanol.

lines joining the various atoms in the molecule. Figure 13 shows an example with methanol, whose formula is CH_3OH.

The method of representation shown in figure 13 is referred to as its 'structural formula'. Practise writing out these formulae with other organic compounds of your choice. Each element exhibits its normal combining power – hydrogen 1, oxygen 2 and carbon 4. Carbon occurs in all organic compounds, oxygen and hydrogen most of the time, and occasionally other non-metals such as nitrogen and sulphur. The existence of the phenomenon of isomerism further increases the huge range of organic compounds that exist.

Organic compounds

In general, organic compounds are covalent in their bonding, although the salts of organic acids – with metals – are ionic in their bonding. Organic compounds have the general characteristics of covalent compounds; they are usually liquids or solids with low melting points. Only those with short carbon chains, fewer than five carbon atoms, are soluble or partially soluble in water. On the other hand they usually mix, or dissolve freely, in organic solvents such as ethanol and benzene. These solutions do not conduct electricity, or else they have very low conductivity.

Functional groups and homologous series

Organic compounds are usually characterised by the presence, in the molecule, of one or more **functional groups.** A single functional group in an organic molecule confers particular properties on it and similar compounds.

- *Example* – The best known example, perhaps is the C–0H group. This occurs in all alcohols. The best known is ethanol, formula C_2H_5OH. Ethanol is only one member of a **homologous series** of alcohols. In this series, as in all others, the member compounds simply increase their carbon chain length by a factor of CH_2.

2	3	4	5
–C–C–	–C–C–C–	–C–C–C–C–	–C–C–C–C–C–

Carbon chains of two to five carbon atoms in length

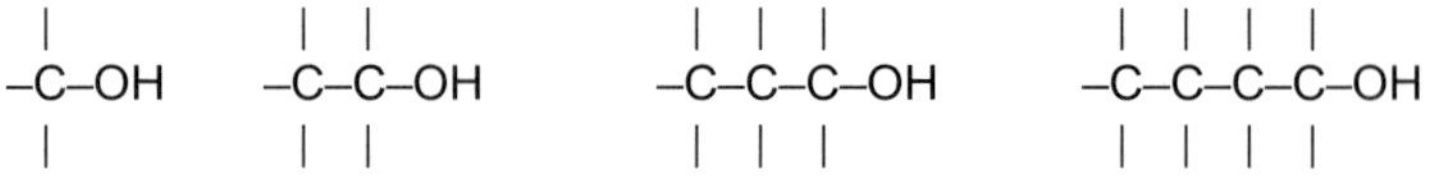

Methanol	Ethanol	Propanol	Butanol
CH_3OH	C_2H_5OH	C_3H_7OH	C_4H_9OH

The remaining bonds all have hydrogen atoms at their ends. Notice the common termination – ol in alcohols. Isomerism in the alcohol series starts with propanol. (For definition see later in this chapter.)

Figure 14. Carbon chains, and the first four members of the alcohol series.

The termination **'ol'** indicates the presence, in each member compound, of the functional – OH group (see figure 14).

Alkanes

The alkanes form another well-known homologous series of formulae, CH_4, C_2H_6, C_3H_8 etc. Again, notice the increase each time by a factor of – CH_2 –. Figure 15 shows a diagram of these three compounds. All compounds containing carbon and hydrogen only are referred to as **hydrocarbons.**

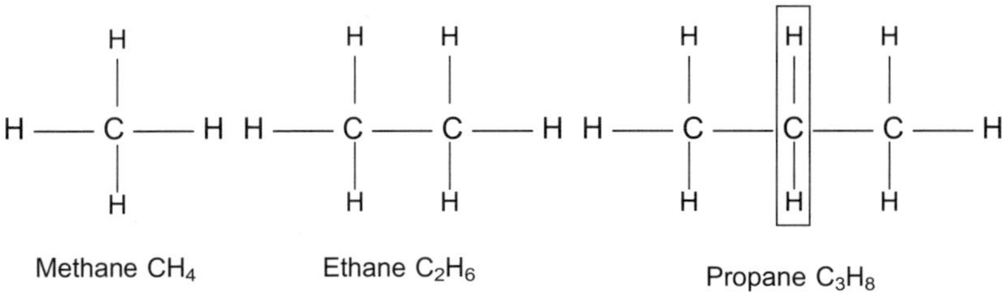

NOTES

1. The alkanes are saturated (contain only single bonds – compare ethene Fig. 16).
2. They increase by an 'increment' of – CH_2 – each time eg see propane above (see also 'homologous series').

Figure 15. The structures of the first three alkanes.

Again, as you can see in figure 15, each member increases by a factor of CH_2. In general in a homologous series the members with the shortest chains have the lowest melting and boiling points. For example, in the above series of alkanes the first four members are gases, but pentane C_5H_{12} is a liquid. The chemical properties of the members of any homologous series are similar – those properties result from the presence of the functional group.

Unsaturated compounds

In the alkanes, the termination 'ane' indicates hydrocarbons which contain carbon atoms with four filled valency bonds. Remember that carbon has a valency (combining power) of four.

► *Key point* – In the homologous series of **alkenes** there is a double bond between a pair of adjacent carbon atoms. See figure 16 for an illustration of this bond.

Compounds that contain a pair of carbon atoms with a double bond between them are said to be **unsaturated.** Despite containing a double bond, unsaturated compounds are more reactive than compounds containing only single bonds.

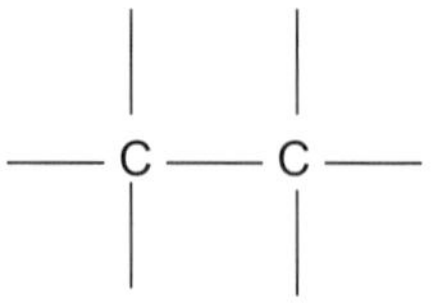

C – C single bond as in ethane and ethanol (Fig. 13)

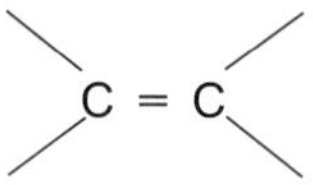

C = C double bond, as in ethene C_2H_4, and propene C_3H_6.

NOTES

1. Compounds containing *single* bonds only are said to be saturated; those with one or more double bonds are described as unsaturated.
2. Unsaturated compounds are generally more reactive than saturated ones.

Figure 16. Carbon-carbon single (C–C) and double (C=C) bonds.

The commonest and simplest alkene is ethene, formula C_2H_4. Its reactivity is due to the fact that it 'wants' to add on two more bonds and become saturated. The structure of ethene is shown in figure 17.

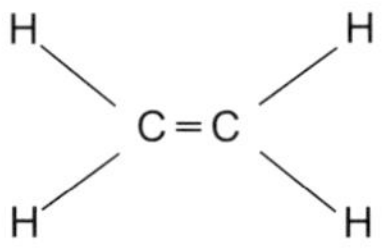

NOTES

1. Ethene is unsaturated (see Fig. 16).
2. It is the first member of the alkenes, general formula C_nH_{2n}.
3. Ethene burns in air with a 'smoky' flame, owing to its relatively high carbon content.

Figure 17. The structural formula of ethene, C_2H_4.

Tests for alkenes

Ethene is a gas, and is produced by the 'cracking' of long chain hydrocarbons. If ethene is passed into **bromine water** – easily made by dissolving a little bromine in cold water in a test tube – the orange red colour of the bromine water disappears. The bromine, an element with formula Br_2, adds on 'across the double bond' producing dibromoethane. The changes that have taken place are shown in figure 18. Ethene is used in the manufacture of polythene and ethanol.

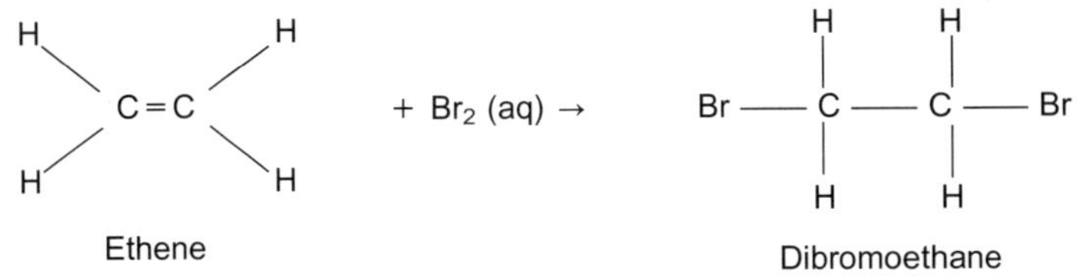

NOTES

1. This reaction is used as a test for unsaturation. The bromine adds 'across' the double bond. This means that one bromine atom attaches to each carbon atom.
2. The double bond has 'disappeared' – so too has the unsaturation. The orange colour is discharged.
3. There is an isomer of dibromoethane where both Br atoms are attached to the same carbon atom.

Figure 18. The reaction between ethene and bromine. (Note – the bromine is normally in the form of bromine water. This is a solution of bromine in water. It is reddish orange in colour.)

Now both carbon atoms have their normal four single bonds, and the unsaturation in ethene has disappeared.

Using acidified **potassium permanganate** *solution for the test*
The colour of this solution is initially purple as the MnO_4^- ion contains manganese with an oxidation number of 7 – that is, not the normal Mn^{++} ion. When an alkene is passed into it, the solution becomes colourless. This is because alkenes, being unsaturated, are reducing in their action.

Fermentation

When yeast, which is a living organism, is added to a sugar solution in a warm (25–30°C) environment, a slow reaction takes place and the sugar is gradually converted into aqueous ethanol and carbon dioxide. The reaction can be shown like this:

$$\text{sugar solution} + \text{yeast} \rightarrow \text{solution of ethanol} + CO_2$$

Fermentation takes place, and the mixture froths as the carbon dioxide is given off. Fermentation stops if the solution becomes too cold, if the ethanol becomes too concentrated, or if all the sugar has been used up. If required, the solution can be filtered to remove the unwanted yeast, and the dilute solution, in water, of the ethanol, can be concentrated by distillation.

► *Note* – It is illegal in the UK to attempt to concentrate aqueous ethanol without a licence from Customs and Excise.

Properties of ethanol

Ethanol is a colourless liquid with a faint pleasant smell and a boiling point of 78°C.

1. Ethanol is fairly reactive – for example it can be fairly easily oxidised to ethanoic (acetic) acid. This is what happens when alcoholic liquids, such as wine, turn sour after prolonged exposure to air.

2. Ethanol burns easily with quite a large heat energy output. If it can be made on a large scale fairly cheaply, it could well replace petrol in car engines.

Ethanol is not normally made on a large scale by fermentation, but from ethene. Ethene and water (steam) react together to form ethanol. The equation is:

$$C_2H_4\ (g)\ +\ H_2O\ (g)\ \rightarrow\ C_2H_5.OH$$

However, fermentation could replace ethene as the starting point for the manufacture of ethanol as oil supplies are used up (ethene comes from crude oil).

Properties of ethanoic (acetic) acid, $CH_3.COOH$

This substance is formed by the oxidation of ethanol. Its structure is shown in figure 19.

The termination '–COOH' is the functional group present in all organic acids.

1. Pure ethanoic acid is a colourless liquid with a sharp smell.

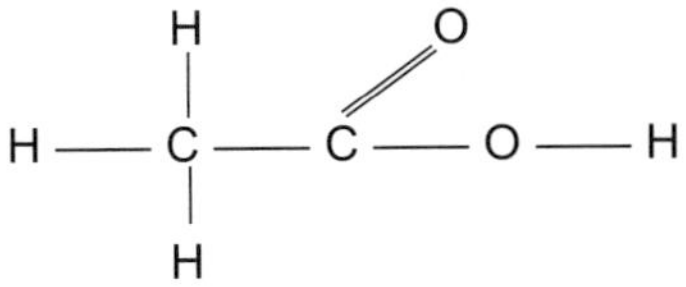

Ethanoic acid CH_3 COOH

NOTES

1. It is better to write out the formula as CH_3. OOH (rather than $C_2H_4O_2$), as the *functional group – COOH is then clearly shown.
2. Ethanoic acid is the second member of a *homologous series of acids. The first is methanoic acid H.COOH.
3. * For an explanation of these terms, see chapter 9.

Figure 19. Structural formula of ethanoic acid.

2. Dilute (aqueous) ethanoic acid is present in vinegar. It is a weak acid. In other words, the concentration of H^+ ions in the aqueous solution is quite low. However, it still gives the basic reactions of an acid. It reddens blue litmus, gives hydrogen with more reactive metals, and carbon dioxide with carbonates.

An additional and important reaction is with ethanol. Under the right conditions the two compounds react together to give ethyl ethanoate which is an example of an ester. The reaction is:

$$\text{ethanol} + \text{ethanoic acid} \rightarrow \text{ethyl ethanoate} + \text{water}$$

The main use of ethanoic acid is in the textile industry, where it is used in the production of man-made fibres. Another use is in the food industry – for example it is present in some salad creams.

Polymerisation

Polymerisation means the joining together of a number of simple molecules of the same kind to form a much larger molecule. For example, ethene polymerises, under pressure, to form poly ('many') ethene, shortened to **polythene**.

Polythene

In the polymerisation process, hundreds of ethene molecules join together to form polythene, the well-known plastic. The general name for this process is **addition**.

$$nC_2H_4 \rightarrow (C_2H_4)_n$$

This important reaction, and the ones discussed earlier with bromine water and permanganate solution, result from the presence of the carbon-carbon double bond. Alkanes, which are saturated, do not give any of these reactions.

The manufacture of ethene

Ethene is manufactured on a very large scale by 'cracking' in the oil industry. In this process, long chain alkanes – often containing up to 20 carbon atoms, which are found in crude oil – are broken down to give a great variety of hydrocarbons with much shorter chains. The shorter chains usually contain from two to eight carbon atoms. Petroleum is largely composed of **octanes**, which are hydrocarbons containing eight carbon atoms. There are several octanes; see the next section on isomerism.

Isomerism

There are several types of isomerism.

- *Key point* – Isomerism occurs when two or more compounds have the same molecular formula, but have different structural formulae.

They are therefore different compounds and have different physical and chemical properties.

Example

Imagine a compound with a chain of three carbon atoms, and any single valent group x at the end, the rest being hydrogen. Its structural formula is therefore:

$$\begin{array}{ccccccc} & | & & | & & | & \\ - & C & - & C & - & C & - X \\ & | & & | & & | & \end{array}$$

where X is a single valent group, say – OH, or atom say – Cl, the remaining bonds on the three carbon atoms all being hydrogen.

There will be an **isomer**, in which X is attached to the *middle* carbon atom in the chain of three. It will therefore be a *different* compound, from that where it is attached to an end carbon atom. Its structure is shown in figure 20.

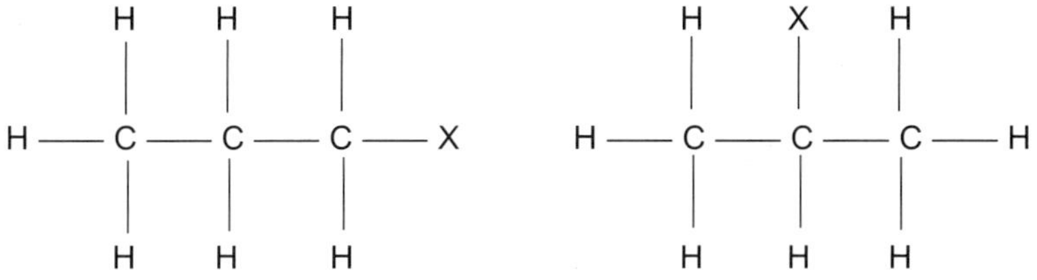

NOTES

1. X is any monovalent atom (eg Cl) or group (eg OH) (see Figure 21). 'Monovalent' means attached to carbon with a single bond.
2. It is possible in the above to substitute more than one hydrogen atom in the molecule. This will give rise to more isomers.

Figure 20. The two isomers in a three carbon chain, with one substituted hydrogen atom.

If X in figure 20 is the grouping –OH, present in all alcohols, the two compounds are as shown in figure 21.

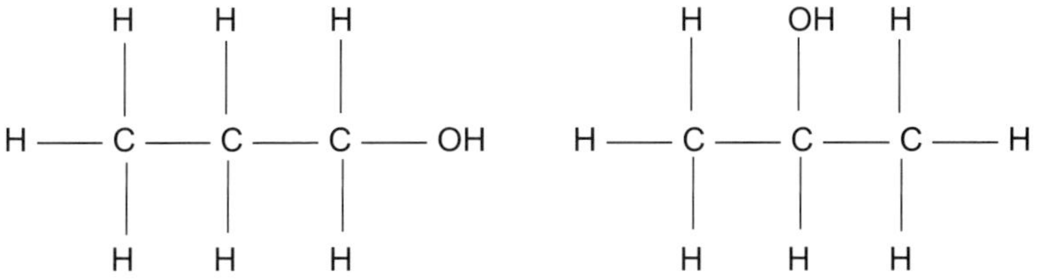

Propan – 1 – ol
C_3H_7OH

Propan – 2 – ol
C_3H_7OH

NOTES

1. In propan-1-ol the -OH group is attached to a terminal carbon atom –in propan-2-ol it is attached to the central carbon atom.
2. The two compounds have slightly different chemical and physical properties – e.g. both are colourless liquids but have different boiling points.

Figure 21. The structural formulae of the two isomeric propanols.

The two compounds shown in figure 21 are called propan-1 -ol and propan-2 -ol and are both alcohols. They are fairly similar, but have slightly different physical and chemical properties; for example their boiling points are not the same.

Their reactions are similar to those of ethanol.

Helping you learn

Progress questions

1. The alcohol butanol, formula $C_4H_9.OH$, has four isomers all containing the functional – OH group.

   (a) Explain clearly the term isomer.

   (b) Explain what is meant by a functional group.

   (c) Write out the structural formulae of the four isomers mentioned above.

2. Explain the difference in structure between an alkane and an alkene.

   (a) Describe a test to distinguish between the two.

   (b) Describe how ethene is converted to polythene. Mention some uses of polythene.

   (c) Write out the structural formula for propane.

   (d) The substance propene has the formula C_3H_6. Write out a structural formula for it.

   (e) There is also an isomer of formula C_3H_6 called cyclopropane. Suggest a structural formula for this compound.

Discussion points

1. How is an aqueous solution of ethanol made by fermentation?

2. How could a fairly pure sample of ethanol be produced from the aqueous solution?

3. Write a balanced equation for the combustion of ethanol.

4. Suggest a possible economic way of the large-scale production of ethanol by the fermentation process.

Key points

1. Organic chemistry is the chemistry of **carbon** and its compounds, many of which occur naturally.

2. The **bonding** between the atoms in organic molecules is **covalent** and they exhibit the general characteristics of covalent compounds.

3. Organic compounds usually contain one, or more, **functional groups** that give rise to their particular properties.

4. A **homologous series** is one where all the compounds contain the same functional group. Each member increases in its chain length by $- CH_2 -$.

5. **Unsaturated compounds** contain at least one **double bond**. This occurs between adjacent pairs of carbon atoms. They are generally more reactive than saturated compounds.

6. **Polymerisation** occurs when large numbers of small molecules containing a double bond join together to form a single much larger molecule. The process is known as addition.

7. **Fermentation** is the conversion of sugar solution to aqueous ethanol by the action of yeast.

8. **Isomerism** occurs when two or more compounds have the same molecular formula, but different structural formulae.

Study and revision tips

1. Organic chemistry is a topic which involves some direct learning. Make sure you understand the terms used in organic chemistry set out in the section above.

2. Exam questions asking you to write out structural formulae of simple organic compounds are quite popular.

10

The Earth and Geological Changes

One-minute summary – Check your own syllabus to see what depth of treatment is required in this topic. In this chapter we discuss the make up of the Earth, the 'rock-cycle', and the valuable source of raw materials provided by the Earth's crust. We also discuss the problems caused by the continuous exploitation of these resources. The Earth is thought to be about 500 million years old, although, on that time scale, life has appeared only very recently. The great bulk of the Earth consists of minerals and rocks in a layered structure. The three main layers are the **core**, the **mantle** and the **crust**. The atmosphere is a part of the upper layer. Organic matter forms a small part only. In this chapter we will explore:

- the rock cycle
- plate tectonics
- the earth as a source of minerals
- pollution.

The rock cycle

The rock cycle is the slow process of changes that occur in the rock structure of the Earth. It involves igneous, metamorphic and sedimentary changes each of which take place over different time scales.

Igneous rocks

Igneous rocks, basalt and granite, are the oldest rocks. They were formed by the cooling of hot molten material called **magma**. They are the result of the hot magma having been forced upwards from the mantle.

Sedimentary rocks

Sedimentary rocks are formed by the gradual process of compression of rock fragments. These rock fragments are the result, over a very long period of time, of weathering. Weathering is the gradual breaking down of large rock masses by fracturing. It is caused

1. by the effects of the expansion of water into ice
2. by the acidic content of rain water, sun and wind
3. by plant growth.

Some weathering also results from the erosion of rocks by the continuous action of both river water and sea water.

Fossils are to be found in sedimentary rocks. These provide evidence about the age of the rock as well as knowledge about the life forms that existed on earth at that time.

► *Examples* – limestone and sandstone.

Metamorphic rocks

Metamorphic rocks are those which result from the effects of high temperatures and pressures on other rocks whilst buried deep underground. The eventual shape, in terms of folding and faulting, of large rock masses is determined by the extent of the forces involved during their formation.

► *Examples* – marble and slate.

Plate tectonics

Plate tectonics refers to the crust of the Earth as a set of huge slow moving plates. It is thought that this movement is caused by the high temperatures underneath the crust. It may be helped by heat generated from natural radioactive decay. This movement is largely responsible for:

1. the recycling of rocks

2. earthquakes

3. volcanic activity

4. the formation of fold mountains.

The Earth as a source of minerals

The Earth is a vast reserve of minerals that human beings have sought to utilise for many hundreds of years. There are numerous examples – salt, phosphates, crude oil and sulphur are just a few.

The problems of consumption

In recent times, doubts have been expressed over the constant plundering of the Earths resources. The supply of resources is dwindling, and there are environmental factors to think about. In some cases it is unclear where long-term substitutes could be found. Environmental pollution is another major cause for concern.

Possible solutions

The long-term solutions to these problems seem to lie in:

1. the better initial use of raw materials

2. recycling

3. the use of substitute materials which make less heavy demands on the environment. In the previous chapter, for example, we mentioned that petrol could perhaps be replaced by ethanol for use in the internal combustion engine.

Pollution

Pollution has become a major source of public concern in recent times. There has been an increase in waste gases in the atmosphere, and in river pollution, caused by the discharge of manufacturing effluent from industrial sites.

1. Atmospheric pollution, caused mainly by sulphur and nitrogen dioxides, is possibly less serious of the two, since the time taken for it to correct itself is less, once the discharge of pollutant gases has been much reduced.

2. Some forms of water pollution take years to disappear. In the meantime, extensive damage is caused to river and marine life.

These problems are now being tackled, as a result of an increasing awareness of how serious they are.

Helping you learn

Progress questions

1. List four important minerals which are mined from the Earth's crust on a large scale. For any *two* of these, describe how they are refined, and state their eventual use.

2. Explain what is meant by the 'fixation' of nitrogen. Describe the essential steps in converting nitrogen to a nitrogenous fertiliser. Like phosphorus, nitrogen is essential to plant growth. Why are the problems of manufacturing nitrogenous fertilisers less serious than those of phosphates?

3. Explain what is meant by the rock cycle. Describe in detail the various stages in the rock cycle.

Discussion points

Phosphorus, in the form of phosphates, is a very important fertiliser for healthy plant growth. The continued mining of rock phosphate has already caused a serious problem for islanders in the southern hemisphere. Can you suggest a solution? Consider two major ways of tackling this problem:

(a) ways of finding alternative sources of phosphate fertiliser, for example by waste control

(b) if the mining has to continue in its present form, ways of limiting environmental damage to the absolute minimum.

Key points

1. The rock cycle involves the continuous processes of rocks being broken down by weathering, whilst the fragments of rocks so formed are compressed into larger masses again. The two take place simultaneously. The three most important groups of rocks are igneous, sedimentary and metamorphic.

2. The Earth is a layered structure consisting of a fairly thin crust on top, a semi-solid mantle, and a very hot core.

3. Large mineral deposits exist in the crust, and man has exploited these in the chemical industry. Concerns are being expressed about the possibility of the supply of these eventually becoming exhausted.

4. Pollution is an additional problem. Better control of waste products, and recycling, may well provide the answers to these problems in the future.

Study and revision tips

1. This chapter is largely one of factual recall, and so will test your memory.

2. Examination questions will not just test your factual knowledge, but your ability to think originally. You may be asked to speculate, for example, on the solution of a particular social or environmental problem. eg, how should scarce resources be allocated? How should the various forms of pollution be dealt with?

11

Industrial Chemistry

One-minute summary – In this final chapter you will learn about some important industrial processes, some of which have already been mentioned. We meet many of these products in our everyday life; others produce chemicals that are essential to other production processes in the chemical industry. Thus we return to the theme of the importance of chemistry in our daily lives. Some products manufactured on a large scale were mentioned in chapter 1. It is a good idea to have a sound working knowledge of the most important of these. In this chapter we will explore:

- chemicals produced from salt and brine
- concentrated acids
- industrial gases
- industry and the environment
- tests for common gases used in the laboratory.

Chemicals produced from salt and brine

The chemicals produced or extracted from salt and brine are principally sodium (metal), sodium hydroxide and chlorine. They were dealt with in detail in chapter 6 (electrolysis).

Concentrated acids

1. Hydrogen chloride and hydrochloric acid (HCl)

Hydrogen gas, from various sources, and chlorine, from the electrolysis of brine, are burned together to produce hydrogen chloride. Hydrogen chloride is a gas.

$$H_2\ (g) + Cl_2\ (g) \rightarrow 2HCl\ (g)$$

This gas is then dissolved in water to give aqueous hydrochloric acid. The 'concentrated' acid is about a 40% solution and it fumes in air. Some of its uses are in the pharmaceuticals industry, in cleaning metals prior to electroplating and as a laboratory reagent.

2. Nitric acid (HNO_3)

Nitric acid is manufactured from ammonia. Its importance lies in the fixation of nitrogen, discussed in chapter 10. This is the conversion of nitrogen from the atmosphere into nitrogen compounds.

The basic reaction in the production process is:

$$\underset{\text{ammonia}}{4NH_3\ (g)} + \underset{\text{oxygen}}{5O_2\ (g)} \rightarrow \underset{\text{nitrogen monoxide}}{4NO\ (g)} + 6H_2O\ (g)$$

The ammonia is obtained via the Haber process, and the oxygen from liquid air. The nitrogen monoxide thus obtained is then fairly easily converted into nitric acid. It is widely used to make nitrates as fertilisers to replace nitrogen lost from the ground. Other uses are to make explosives and in the laboratory.

The pure acid is a nearly colourless, very corrosive liquid, which fumes in air. The salts it forms are called nitrates. It is a powerful oxidising agent.

3. Sulphuric acid (H_2SO_4)

Sulphuric acid is manufactured by the 'contact' process described in chapter 7. Pure sulphur dioxide, from various sources, and oxygen, from liquid air are passed together over a heated catalyst (500°C) to produce sulphur trioxide (SO_3):

$$2SO_2\ (g) + O_2\ (g) \rightarrow 2SO_3\ (g)$$

The sulphur trioxide is then converted into sulphuric acid. Pure starting gases are essential, as catalytic 'poisoning' takes place if impurities are present. In addition, an impure product results, that is almost impossible to purify.

Other important substances produced by the chemical industry include sodium carbonate and bicarbonate. Check with your syllabus.

Industrial gases

Hydrogen, some oxygen, and chlorine

These gases are produced from the electrolysis of sodium chloride (both molten and brine). Full details are given in chapter 6. Some hydrogen is produced from natural gas.

Carbon dioxide (CO_2)

This gas was once made by heating limestone (calcium carbonate). Most carbon dioxide is now made as a by-product of fermentation. This is discussed in chapter 9. Carbon dioxide is used in making fizzy drinks, in fire extinguishers, and in hospitals.

Ammonia (NH_3) – the Haber process

Nitrogen, from liquid air, and hydrogen gas are mixed in a ratio of one to three by volume at 200 atmospheres pressure and 500°C. They are then passed over a heated catalyst:

$$N_2\ (g) + 3H_2\ (g) \rightarrow 2NH_3\ (g)$$

The gases must be pure, otherwise the catalyst is 'poisoned'. In other words its efficiency is reduced. The reaction is reversible and the conditions used are chosen to promote the formation of ammonia. Ammonia, which is a gas, is used in the manufacture of:

1. nitrogenous fertilisers
2. household cleansing agents (as a solution in water)
3. ammonium salts
4. nitric acid
5. nylon.

Nitrogen (N_2)

Nitrogen is made from liquid air by fractional distillation. It is used to make ammonia in the Haber process, and to store organs in deep freeze prior to their use in surgery.

Oxygen (O_2)

Oxygen is obtained from liquid air, along with nitrogen. A limited amount is obtained from electrolysis. It is used in hospitals, and to supplement the natural supply if this is restricted, and in metal cutting and welding.

Industry and the environment

Some effects of industry on the environment were discussed in the last chapter. There has been a big increase in people's awareness of industrial pollution caused by the chemical industry, especially the pollution of air and water. Much stricter control of effluent discharge is now required by the government. There are substantial fines for manufacturing organisations that do not 'clean up their act'.

Tests for common gases used in the laboratory

Why has this topic has been included in this chapter? The reason is that gases play a big part in the chemical industry. Some of them are intermediates in producing other important compounds. Examples are ammonia, sulphur dioxide and ethene. Table 13 below outlines the main properties and tests for all the important gases in chemistry.

Gas	Appearance/smell	Test
hydrogen	colourless, no smell	Lighted splint ignites gas with a squeak or pop.
oxygen	colourless, no smell	A glowing splint re-lights.
nitrogen	colourless, no smell	Negative to tests.
chlorine	green, pungent smell	Reddens, and then bleaches moist blue litmus paper.
sulphur dioxide	colourless, choking smell	Turns moist blue litmus paper red.
hydrogen chloride	colourless, fumes, acrid smell	Gives white fumes with the stopper of an ammonia bottle.
ammonia	colourless, pungent characteristic odour	Turns moist red litmus paper blue.
nitrogen dioxide	brown, smells like nitric acid	Turns moist blue litmus paper red.
carbon dioxide	colourless, very little smell	Lime water turns milky.
carbon monoxide	colourless, no smell, toxic	Burns with a flickering blue flame at the mouth of a test tube

Table 13. Tests for common gases.

Helping you learn

Progress questions

1. You are asked to prepare some nitrogen dioxide (NO_2), which is a gas, in the laboratory. It can be generated fairly easily by heating lead nitrate, $Pb(NO_3)_2$. The gas produced is soluble in water.

   (a) Sketch a suitable apparatus for generating and collecting some of the gas, bearing in mind that oxygen is also produced.

   (b) Write a balanced equation showing the decomposition.

   (c) Why is lead nitrate chosen and not, say, zinc nitrate?

2. Write an account of how you see a possible solution to the conflict of interests, over an extended time span, between the mining industry and environmental interests.

Key points

'Green' issues have come to the fore in recent years. A conflict has arisen between parts of the chemical industry and environmentalists. Pollution has become a fashionable talking point, but may be capable of resolution; some of its effects are reversible. Irreversible changes are far more serious, and deserve more attention from environmentalists. The main factors are the road and house building industries, and their natural precursor the mining industry. It is hard to see a long-term solution. It is certainly a discussion point, as quarrying has already resulted in protests from countryside lovers, and no doubt will continue to do so.

Practical assignments

It is well worth generating each one of the gases described in table 13, in a test tube, and collecting a sample. As you do this, note in each case how it is made on a larger scale in the laboratory. Do not do this for carbon monoxide (which is poisonous), and take particular care with chlorine (which is toxic). In the case of carbon monoxide see if you can find out how to make some on a test tube scale. Apply the tests shown in table 13 to the gases.

Study and revision tips

1. The properties and tests for the gases given in table 12 at the end of this chapter should be committed to memory.

2. Make sure you can talk and write about the large-scale industrial production of oxygen, carbon dioxide and ammonia, and the manufacture of the three acids (hydrochloric, nitric and sulphuric).

3. You may be tested on your knowledge of the main causes of pollution of air and drinking water supplies, and the measures being taken to control them. You may be asked to say whether you think these measures are adequate, and what else might be done, for example by government legislation.

Appendix 1: Table of Atomic masses

Note: only common elements are included.

Element	Symbol	Approximate atomic mass
aluminium	Al	27
argon	Ar	40
barium	Ba	137
bromine	Br	80
calcium	Ca	40
carbon	C	12.000 (Standard used)
chlorine	Cl	35.5
copper	Cu	63.5
fluorine	F	19
helium	He	4
hydrogen	H	1
iodine	I	127
iron	Fe	56
lead	Pb	208
magnesium	Mg	24
manganese	Mn	55
mercury	Hg	201
nitrogen	N	14
oxygen	O	16
phosphorus	P	31
potassium	K	39
silver	Ag	108
sodium	Na	23
strontium	Sr	88
sulphur	S	32
tin	Sn	119
zinc	Zn	65

Group																	
I	II											III	IV	V	VI	VII	O
							H Hydrogen 1										He Helium 2
L1 Lithium 3	Be Beryllium 4											B Boron 5	C Carbon 6	N Nitrogen 7	O Oxygen 8	F Fluorine 9	Ne Neon 10
Na Sodium 11	Mg Magnesium 12											Al Aluminium 13	Si Silicon 14	P Phosphorus 15	S Sulphur 16	Cl Chlorine 17	Ar Argon 18
K Potassium 19	Ca Calcium 20	Sc Scandium 21	Ti Titanium 22	V Vanadium 23	Cr Chromium 24	Mn Manganese 25	Fe Iron 26	Ni Nickel 28	Co Cobalt 27	Cu Copper 29	Zn Zinc 30	Ga Gallium 31	Ge Germanium 32	As Arsenic 33	Se Selenium 34	Br Bromine 35	Kr Krypton 36
Rb Rubidium 37	Sr Strontium 38	Y Yttrium 39	Zr Zirconium 40	Nb Niobium 41	Mo Molybdenum 42	Tc Technetium 43	Ru Ruthenium 44	Rh Rhodium 45	Pd Palladium 46	Ag Silver 47	Cd Cadmium 48	in Indium 49	Sn Tin 50	Sb Antimony 51	Te Tellurium 52	I Iodine 53	Xe Xenon 54
Cs Casium 55	Ba Barium 56	La Lanthanum 57	Hf Hafnium 72	Ta Tantalum 73	W Tungsten 74	Re Rhenium 75	Os Osmium 76	Ir Iridium 77	Pt Platinum 78	Au Gold 79	Hg Mercury 80	Tl Thallium 81	Pb Lead 82	Bi Bismuth 83	Po Polonium 84	At Astatine 85	Rn Radon 86
Fr Francium 87	Ra Radium 88	Ac Actinium 89															
				Ce Cerium 58	Pr Praseody'm 59	Nd Neodymium 60	Pm Promethium 61	Sm Samarium 62	Eu Europium 63	Gd Gadolinium 64	Tb Terbium 65	Dy Dysprosium 66	Ho Holmium 67	Er Erbium 68	Tm Thulium 69	Yb Ytterbium 70	Lu Lutetium 71
				Th Thorium 90	Pa Protoactin'm 91	U Uranium 92	Np Neptunium 93	Pu Plutonium 94	Am Americium 95	Cm Curium 96	Bk Berkelium 97	Cf Californium 98	Es Einsteinium 99	Fm Fermium 100	Md Mendelevi'm 101	No Nobelium 102	Lr Lawrencium 103

Periodic Table of the Elements

Answers to Progress Questions

Chapter 1

1. Salt and iron filings. A fairly good separation can be achieved by the careful use of a suitable magnet. A second method involves using hot water added to the mixture. The salt will dissolve and the mixture poured through a filter paper in a funnel. The iron filings will be retained in the filter paper whilst the salt solution will pass through.

2. Industrially the two gases are separated by the fractional distillation of liquid air. This method has to be used as the boiling points of the two are close together. In the laboratory the oxygen can be removed by passing air over a suitable heated metal such as iron filings. The nitrogen is not absorbed as it does not react with oxygen.

3. The water in sea water may be obtained by ordinary distillation using a water cooled condenser. The salt in sea water is left in the distillation flask.

Chapter 2

2. (a) Br_2, 2Zn, SO_2, HCl

3. Positive ions K (K^+), Cu (Cu^{++}), H (H^+), Ba (Ba^{++}), Ag (Ag^+)

   Negative ions S (S^{--}), Br (Br^-), O (O^{--})

   Note that hydrogen occasionally forms negative ions.
   Metals form positive ions, non-metals form negative ions.

4. The inert gases contain a full outer shell of eight electrons, and can neither gain or lose them.

Chapter 3

1. Tests should be a combination of physical and chemical tests. Their melting points and solubility in water will indicate the nature of both. Ionic substances generally have high melting points and are soluble in water. In addition the pH of the solutions can be measured. Non-metallic oxides and chlorides usually give acidic solutions with water. The electrical conductivity of the molten solid will indicate the nature of its bonding. If the conduction is poor the bonding is covalent, or partially so, and the chloride will be a non-metallic one.

2. Strontium lies between calcium and barium in Group II. Since reactivity, in groups of metallic elements increases downwards it will be more reactive towards water and acids than calcium, but less so than barium. Its oxide will be a strong base, and can be made by the thermal decomposition of either its nitrate or carbonate, but a high temperature will be required. The salts will be ionic in their bonding. The metal will require an electrolytic extraction method.

3. The element astatine, at the bottom of Group VII will fairly closely resemble iodine. It will be a solid, with a fairly low melting point and boiling point. It will be the least reactive of the halogens, but will probably form the At^- ion with more reactive metals. This ion will very easily lose its charge, reforming the element.

Chapter 4

1. $4NH_3\ (g) + 5O_2\ (g) \rightarrow 4NO_2\ (g) + 6H_2O\ (g)$

2. The molar mass of lead nitrate is:

$$207 + 2(14 + 3(16)) = 207 + 2(62) = 207 + 124 = 331\ g$$

3. The molar mass of hydrated copper sulphate is:

$$+ 32 + 4(16) + 5(2+16) = 63.5 + 32 + 64 + 90 = 249.5g$$

4. Formula is $NaHCO_3$, and the formula mass is 84g.

$$\% \text{ of sodium is } \frac{23}{84} \times 100 = 27.38\ \%$$

$$\% \text{ of hydrogen is } \frac{1}{84} \times 100 = 1.19\ \%$$

$$\% \text{ of carbon is } \frac{12}{84} \times 100 = 14.29\ \%$$

$$\% \text{ of oxygen is } \frac{48}{84} \times 100 = 57.14\ \%$$

5. From the equation for the reaction, 1 mole zinc gives 1 mole of hydrogen, ie 65g Zinc gives 22.4 dm^3 of hydrogen.

$$1\text{g gives } \frac{22.4}{65}\ dm^3 \text{ and 5g gives } 22.4 \times \frac{5}{22.4} = 1.723 dm^3$$

6. The ratio of moles from the equation is 1.
Therefore 27.5 x M = 25 x 0.095 (M is the unknown)
From which the molarity of the nitric acid was 0.086.

7. The balanced equation, with state symbols is:

$$2Pb(NO_3)_2\ (s) \rightarrow 2PbO\ (s) + 4NO_2\ (g) + O_2\ (g)$$

Chapter 6

1. 47.4g of copper. Did you remember that 2 Faraday are required to discharge a mole of copper?

2. $MnO_4^-(aq) + 8H^+(aq) + 5Fe^{++}(aq) \rightarrow Mn^{++}(aq) + 4H_2O + 5Fe^{+++}(aq)$

Chapter 7

1. The equation for the action of heat on hydrated magnesium nitrate is:

$$2Mg(NO_3)_2.6H_2O \rightarrow 2MgO\ (s) + 4NO_2(g) + O_2(g) + 12H_2O$$

The salt initially dissolves in its own water of crystallisation. It then becomes a solid and gives off both nitrogen dioxide and oxygen. White magnesium oxide is left at the end.

2. Barium chloride solution.

3. When lead iodide is boiled the precipitate re-dissolves, forming aqueous lead and iodide ions which are both colourless.

4. (a) Precipitation and a reversible reaction.
   (b) Oxidation and reduction since the numbers of charges on the ions has changed.
   (c) Oxidation and reduction, and precipitation.
   (d) Thermal decomposition.

Answers to discussion point questions

1. There will be a reduction in volume, from approximately 3 volumes, since there are 3 moles of reactants, to 2 volumes (2 moles of products).

2. As there is a reduction in volume the reaction will be favoured by assisting this natural reduction – therefore an increase in pressure will promote the formation of C.

3. As the reaction is exothermic i.e. heat is produced, a high temperature will tend to drive the reaction backwards – i.e. from right to left. The process should therefore be conducted at a fairly low temperature.

4. After the reaction is complete, especially if it is continuous and product C leaves the reaction chamber, there could be a blockage problem in any pipework conducting C away. Therefore, it may be necessary to keep this pipework at a temperature above 30°C all the time.

5. The term 'finely divided' means that the catalyst is employed in some form where the largest surface area is presented to the reacting mixture. This would involve using the metal in the form of mesh, or powder supported on an inert solid.

6. In the event of reduced efficiency, the catalyst can be replaced

with an alternative, as often more than one catalyst is effective. However, the best way to ensure a good life span for a catalyst, is to start with pure starting materials, i.e A and B. This has the added advantage of giving a purer product.

Chapter 8

1. A few chlorine molecules (Cl_2) are dissociated into single atoms by the energy in sunlight. These atoms, with their extra energy, rupture H-H bonds in H_2 molecules, and then react producing hydrogen chloride and extra energy. A chain reaction follows at once.

2. (a) Fast. (b) Moderate. (c) Fast. (d) Slow. (e) Fast.

3. (a) About 8.5 seconds. (b) About 63 cm^3. (c) About 66 cm^3.

4. All the magnesium has been used up.

5. An initially steeper graph than the one shown but with the same maximum volume of hydrogen eventually obtained.

6. Starting with less concentrated acid, and less magnesium, so that the initial rate is less and the final volume of hydrogen produced is less.

Chapter 9

1. (c) See figure 22.
2. (c) $2C_2H_5.OH\ (l) + 7O_2\ (g) \rightarrow 4CO_2\ (g) + 6H_2O\ (l)$
3. (c) See figure 23 (on page 128).

Chapter 10

1. Examples are limestone, sodium nitrate, sulphur, and crude oil. Possible uses are in the petrochemical industry, the building industry, manufacture of 'heavy' chemicals, and the pharmaceutical industry.

There are two 'straight' chain isomers (4 carbon atoms in a row) and also two 'branched' chain (3 carbons in a row) isomers.

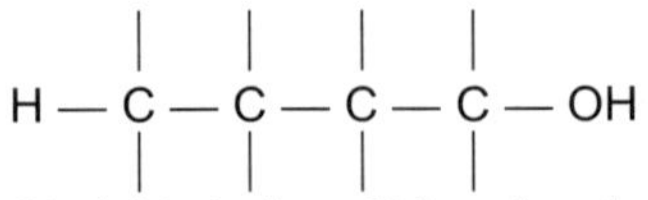

Straight chain – OH on terminal carbon atom

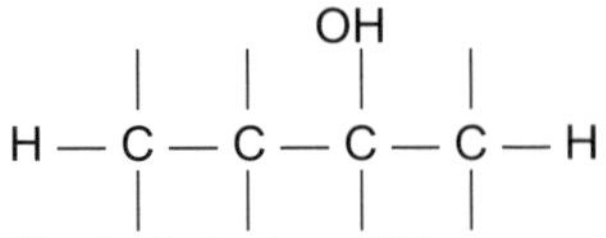

Straight chain – OH on second carbon atom

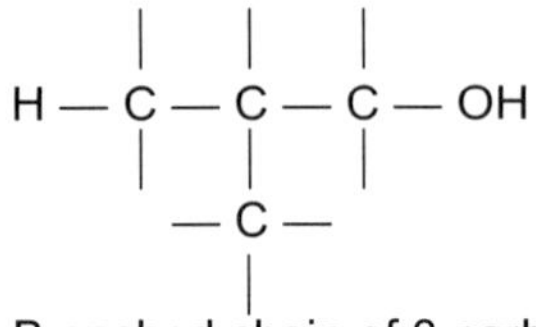

Branched chain of 3 carbon atoms, with -OH group attached to terminal C atom.

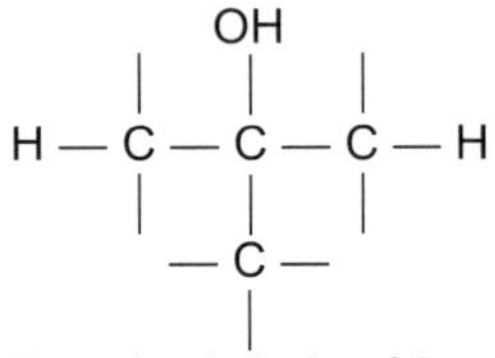

Branched chain of 3 carbon atoms, with -OH group attached to central C atom in chain.

NOTES

1. All the remaining bonds have hydrogen atoms at their ends.
2. Like the 2-propanols, the 4 butanols have fairly similar properties.

Figure 22. The four isomeric butanols, formula C_4H_9OH.

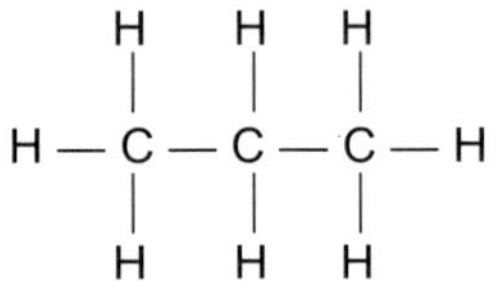

Propane C_3H_8

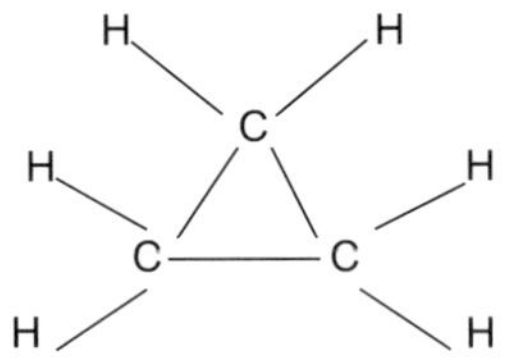

Cyclopropane C_3H_6

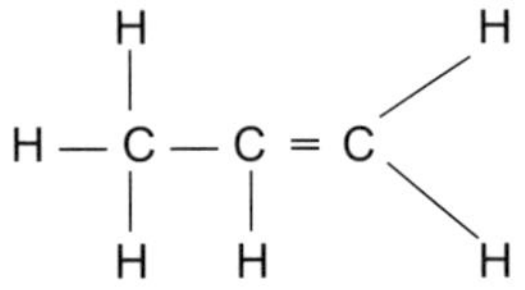

Propene C_3H_6

Figure 23. Structures of propane C_3H_8, cyclopropane C_3H_6, and propene C_3H_6.

2. The term 'fixation of nitrogen' refers to the conversion of nitrogen in the atmosphere to useful nitrogen compounds. The essential steps in the fixing of nitrogen are its initial conversion to ammonia (the Haber process), the conversion of ammonia into nitrogen monoxide, and the conversion of nitrogen monoxide into nitric acid. Most compounds containing fixed nitrogen, such as nitrogenous fertilisers, are made from nitric acid. The reason that the problem of manufacture of nitrogenous fertilisers poses a less serious problem than the manufacture of phosphate fertilisers, is that there is a huge supply of atmospheric nitrogen available. Additionally, most of the nitrogen removed from the atmosphere is eventually returned to it by way of the nitrogen cycle. There is no parallel large supply of phosphate available for exploitation, and only a very minor 'phosphorus cycle'.

3. The rock cycle means the continuous breaking down of rocks by a range of environmental factors, and the simultaneous reformation of rocks by compression in the Earth's crust from fragmentary rocks.

Chapter 11

1. (a) The apparatus should include a fairly large combustion tube, leading to a U-tube. The lead nitrate is heated in the combustion tube. The U-tube is immersed in a freezing mixture (crushed ice and salt mixture) contained in a large beaker. This condenses the nitrogen dioxide to a liquid whilst the oxygen, which is also produced, is not affected.

   (b) $2Pb(NO_3)_2\ (s) \rightarrow 2PbO\ (s) + 4NO_2\ (g) + O_2\ (g)$

   (c) Lead nitrate is normally used in this preparation because it is anhydrous. Most other heavy metal nitrates, including zinc nitrate, contain water of crystallisation. This makes them unsuitable in the experiment, although they also produce nitrogen dioxide as with lead nitrate.

Glossary

acid, acidic (adj.)

alkali, alkaline (adj.)

alkane A saturated (single bonds only) hydrocarbon, which form a homologous series of general formula CnH_{2n+2}. They are stable and unreactive.

alkene A homologous series of unsaturated (containing at least one double bond) hydrocarbons of general formula CnH_{2n}. They are more reactive than alkanes.

alpha particles Helium nuclei.

anhydrous 'Without water' e.g. lead nitrate $Pb\,(NO_3)_2$ is anhydrous, but magnesium nitrate, formula $Mg(NO_3)_2.6H_2O$. The $6H_2O$ is referred to as water of crystallisation.

anion A negative ion.

anode The connection to the positive terminal of the power supply. See also cathode.

aqueous Containing water, or a solution of any substance in water.

atom The smallest particle of an element which can exist on its own, and which can take part in a chemical reaction.

atomic mass number The atomic mass number of an atom is equal to the total number of protons and neutrons added together.

atomic number The atomic number of an element is equal to the number of electrons, or protons since these are the same.

beta particles Fast electrons emitted from the nucleus.

bonds, chemical The links which bind together atoms in the elements and compounds in chemistry.

burette Used in titrations. It consists of a long glass tube clamped vertically, with accurate graduations marked on its side in cm^3. At the bottom end is a glass tap to deliver precise volumes of the liquid inside it.

carbon A non-metallic element, atomic number 6, at the head of

Group IV in the Periodic Table. The '12' isotope is used as the standard of atomic masses.

catalyst A substance which is used to promote (usually) a chemical reaction, by enabling it to proceed at a lower temperature and/or pressure. They do not normally take part in the reaction and can thus theoretically be reused.

cathode The connection to the negative terminal of the power supply. See also anode.

cation A positive ion.

combustion Burning. Combustion takes place when any flammable material burns in oxygen (or air).

compound A single substance made up of two or more elements. See also mixture.

covalent bonds Bonds where one or more electrons from each atom form a shared electron pair situated between the nuclei of the two atoms at either end of the bond.

cracking A process, used in the petrochemical industry, to convert long chain hydrocarbons into shorter chain lower boiling ones. These are principally octanes for use in petroleum.

crystallisation The process of crystal production from a solution of a solid substance (solute) dissolved in a solvent. It can occur naturally, or hastened by heating to drive off the solvent.

distillation The process of heating a solution containing a mixture of more than one substance, to drive off the more volatile one, i.e. with the lower boiling point. It may be used for both separation and purification.

ductility The property of being able to be drawn into a wire.

electrolyte This is a substance in the liquid state – it may be molten, or as a solution in water, which conducts an electric current and at the same time is decomposed by it.

electrolysis The process of passing an electric current (direct) through an electrolyte and decomposing that electrolyte.

elements Substances that can, and very often do, exist as a separate chemical species. They are all composed of atoms, or in a few cases molecules containing two atoms.

endothermic reactions In endothermic reactions heat is absorbed, and the products have more energy than the starting substances.

ethylene An alternative name for ethene C_2H_4.

exothermic reactions In exothermic reactions heat is given out, new bonds are formed and the products have less energy than the starting substances.

fermentation The conversion of sugar solution to aqueous ethanol by the action of yeast.

ferrous The Iron (II) oxidation state of iron – i.e. containing the Fe^{++} ion e.g. as in Ferrous Chloride $FeCl_2$.

filtration The process of separating a solid from a liquid.

formula The representation by means of atomic symbols of all the atoms of all the elements in a substance, e.g. H_2SO_4 shows the elements present and how many atoms of each.

fractional distillation A superior distillation process used to separate several components of a liquid mixture, according to their different boiling points, and achieving a better separation than ordinary distillation.

galvanising The application of a layer of zinc to prevent rusting.

gamma rays Radioactive radiation of very short wavelength.

gas The 'third state' of matter, in which it has no shape, and where the atoms of molecules are free to move around at random. It also fills up all the available space in its container.

geology The study of all the various materials and substances that make up the Earth's structure, i.e. rocks, minerals, oceans, etc.

gram The most commonly used unit of mass in chemistry. The international standard of mass is the kilogram. The gram is one-thousandth of a kilogram.

Haber process The industrial process used for the large-scale production of the gas ammonia, NH_3, from nitrogen and hydrogen. Named after the German chemist Fritz Haber.

hydrocarbons All compounds containing carbon and hydrogen only.

ionic equation An equation limited to those reactions that take place in aqueous solution.

ion A charged particle.

isomerism This occurs when two or more compounds have the same molecular formula, but different structural formulae.

isotope This is an atom(s) of an element of different mass number to other atoms of the element. The existence of isotopes is caused by the presence of extra neutrons in the nucleus, making it relatively heavier or lighter than other isotopes of the same element. The number of protons and electrons remains the same.

liquid The state of matter between solid and gas. Liquids can be poured and will take up the shape of their container.

magma Hot molten material beneath the Earth's crust.

malleability The property of being able to be beaten into a sheet. See also ductility.

mass The term used in chemistry to denote the amount of material in a substance. The unit normally used to measure the amount of a substance is the gram. See also the 'mole'.

metal, metallic Generally hard, durable substances which make up the bulk of the elements in the Periodic Table. They are resistant to stress, have high densitites and are good conductors of heat and electricity. They form positive ions.

metamorphic rocks Those which result from the effects of high temperatures and pressures on other rocks whilst buried deep underground.

mineral This is the name given to the very large range of naturally occuring inorganic substances which exist in the Earth's crust. They are similar to ores but may have a higher natural purity. Oxides, sulphides and halides are examples of minerals.

miscible Mixable.

mixture Two or more substances, not combined with each other, which are normally fairly easy to separate. See also compound.

molarity A way of expressing the concentration of a solution.

mole The formula mass in grams.

molecule A group of at least two atoms, chemically joined together, and which can exist separately.

octanes Hydrocarbons containing eight carbon atoms.

oxidation number Refers to the number of charges – positive or negative or zero – on an atom, an ion or the elements in a compound.

paper chromatography A process used to separate two or more liquids mixed together.

particles Elements, atoms, molecules, ions and electrons

Periodic Table This is an arrangement of all the known elements, using their symbols, in order of their ascending Atomic Number. They are additionally arranged into groups (vertical families) and periods (horizontal groups).

pipette Used in titrations, and consists of a central glass bulb with two glass tubes attached on either side. It is used to accurately measure out and deliver a known volume of liquid, commonly $25cm^3$, of a solution which is to be titrated. The solution is normally introduced into a pipette by suction from a container of that solution.

plate tectonics The term used to describe the crust of the earth as a set of huge slow moving plates.

polymerisation The joining together of a number of simple molecules of the same kind to form a much larger molecule.

polythene A plastic produced from ethene C_2H_4 by polymerisation.

radicals Groups of negatively charged atoms, or anions.

radioactivity The term used to describe the emission of radiation by some elements.

reaction The term used to describe any chemical change taking place between two or more substances (the reactants) resulting in the production of new substances (products).

rock cycle The slow process of changes that occur in the rock structure of the Earth.

sedimentary rocks Rocks formed by the gradual process of compression of rock fragments.

solid The first state of matter. Solids have a definite shape which they retain unless stressed or deformed.

solution Any liquid which contains a solvent and one or more substances (solute(s)) dissolved in that solvent.

standard solution A solution of accurately known concentration.

state There are three states in which substances may exist: solids, liquids, or gases.

stp Standard temperature and pressure.

structural formula Shows all the atoms present in the molecule of that substance, fully written out to show all the individual bonds in that molecule. Normally applied to organic compounds.

sublime, to To turn into vapour at once.

substance A single species, an element or a pure chemical compound.

synthesis The making of a compound, or compounds, from simpler starting materials.

titration Volumetric analysis.

valency The combining power of an element.

water voltameter A special piece of apparatus used to show the electrolysis of water.

Web Sites for Chemistry Students

One-minute summary – The internet, or world wide web, is an amazingly useful resource, giving the student nearly free and almost immediate information on any topic. Ignore this vast and valuable store of materials at your peril! The following list of web sites may be helpful for you. Please note that neither the author nor the publisher is responsible for content or opinions expressed on the sites listed, which are simply intended to offer starting points for students. Also, please remember that the internet is a fast-evolving environment, and links may come and go. If you have some favourite sites you would like to see mentioned in future editions of this book, please write to David Howard c/o Studymates (address on back cover), or email him at the address shown below. You will find a free selection of useful and readymade student links for chemistry and other subjects at the Studymates web site. Happy surfing!

Studymates web site: http://www.studymates.co.uk
David Howard email: davidhoward@studymates.co.uk

Chemistry

ChemExper
http://www.chemexper.com/
This is an database of chemicals. You can explore more than 60,000 compounds and find out their physical properties such as boiling point, melting point and density.

ChemFinder
http://www.chemfinder.com/
ChemFinder indexes hundreds of web sites that contain chemical data.

Chem 4 Kids
http://www.chem4kids.com/
This US site introduces chemistry basics such as atoms, elements and matter.

Chemistry (Salters)
http://www.ocr.org.uk/schemes/gcse/sci/17912000/1791s.htm
This is a page of Oxford Cambridge and RSA Examinations (OCR). It tells you all about the OCR syllabus, coursework and assessment.

CheMystery
http://hyperion.advanced.org/3659/
This is a 'virtual chemistry textbook', offering an interactive guide for high school chemistry students, and links to other internet resources on the world wide web.

Creative Chemistry
http://www.creative-chemistry.org.uk/
This UK site was developed at Harrogate Granby High School. It offers worksheets and teaching notes for fun activities, suitable for a Chemistry Club and GCSE Chemistry. There are interactive revision quizzes and puzzles to help you improve your knowledge of chemistry. Well worth a look.

Links for Chemists (University of Liverpool)
http://www.liv.ac.uk/Chemistry/Links/links.html
This excellent site contains links to over 7,500 chemistry resources on the world wide web, including to UK university chemistry departments. Well worth a look.

Royal Society of Chemistry
http://www.chemsoc.org/indexliv.htm
The leading UK professional and scientific body for chemists.

Sambals Science Web
http://www.sambal.co.uk/chemistry.html
These pages will help you with your Key Stage 3 Chemistry. You may also find them useful if you are doing Chemistry GCSE.

School Chemistry Links
http://ourworld.compuserve.com/homepages/RayLec/chlinks.htm
A useful set of UK and international links for chemistry learning and reference.

Schoolsco
http://www.schools.co.uk/Chemistry.htm
The page contains links to lots of GCSE Chemistry textbooks and revision guides.

Sheffield's Chemdex
http://www.shef.ac.uk/~chem/chemdex/
Since 1993, Chemdex has maintained an international directory of chemistry on the internet. It contains more than 4,000 links, and is run from the University of Sheffield.

Search engines

Internet search engines have become big web sites in their own right. As well as search boxes, you will find handy directories of information, news, email and other services. There are thousands of search engines freely available, but the biggest and best known are probably Alta Vista, Excite, Infoseek, Lycos and Yahoo!. Try out several and see which one(s) you like the best.

AltaVista
http://dir.altavista.com/Science/Chemistry/Organic.shtml
Alta Vista is one of the most popular search sites among web users everywhere. This is its portal to organic chemistry on the web.

Excite
http://www.excite.co.uk/
Excite is another of the top search engines and directories, with separate home pages for several different countries, including this one for the UK.

Google
http://www.google.com
This simple-looking but powerful search engine is proving popular with regular users. A basic keyword search for 'organic chemistry' yielded more than 46,000 results.

Infoseek (Go Network)
http://infoseek.go.com/
Infoseek is another of the top ten search engines on the internet.

Lycos
http://www.lycos.co.uk/
Lycos is another of the top ten worldwide web search engines. This is its UK home page.

Metaplus
http://www.metaplus.com/uk.html
Metaplus is a metalist of the best internet directories – and also offers direct links to some key general sites. This is its UK page, containing hundreds of classifications to explore.

Webcrawler
http://webcrawler.com/
Webcrawler is a fast worker and returns an impressive list of links, locating key words that may be buried deep within a document's text.

WebFerret
http://www.ferretsoft.com
WebFerret is not a search engine as such, but a handy little search utility. It queries ten or more leading search engines all at once, discarding duplicate results. You key in your query offline, and when you connect it searches the web until it has collected the number of references you have specified – up to 9,999 if you wish. The program is free and only takes a few minutes to download. Highly recommended.

Yahoo! UK
http://www.yahoo.co.uk
Yahoo! was the first big search site on the internet, and is still one of the best for free general searching. Millions of people use it every day.

Index